The Building Process
SIMPLIFIED

A Homeowner's and Contractor's Guide to Codes, Permits, and Inspections

The Building Process
SIMPLIFIED

A Homeowner's and Contractor's Guide to Codes, Permits, and Inspections

Linda Pieczynski

DELMAR
CENGAGE Learning

Australia • Brazil • Japan • Korea • Mexico • Singapore • Spain • United Kingdom • United States

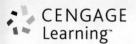

**The Building Process Simplified:
A Homeowner's and Contractor's Guide to
Codes, Permits, and Inspections**
Author: Linda Pieczynski

Vice President, Technology and Trade
    Professional Business Unit: Gregory L.
    Clayton
Director of Learning Solutions:
Product Development Manager: Ed Francis
Product Manager: Vanessa Myers
Editorial Assistant: Nobina Chakraborti
Director of Marketing: Beth A. Lutz
Executive Marketing Manager: Taryn Zlatin
Marketing Manager: Marissa Maiella
Production Director: Carolyn Miller
Production Manager: Andrew Crouth
Content Project Manager: Brooke
    Greenhouse
Art Director: Benjamin Gleeksman

Library of Congress Control Number: 2008941270

ISBN-13: 978-1-4354-2847-8
ISBN-10: 1-4354-2847-1

Delmar
5 Maxwell Drive
Clifton Park, NY 12065-2919
USA

Cengage Learning is a leading provider of customized learning solutions with office locations around the globe, including Singapore, the United Kingdom, Australia, Mexico, Brazil, and Japan. Locate your local office at: **international.cengage.com/region**

Cengage Learning products are represented in Canada by Nelson Education, Ltd.

To learn more about Delmar, visit **www.cengage.com/delmar**
Purchase any of our products at your local college store or at our preferred online store **www.ichapters.com**

NOTICE TO THE READER

Printed in Canada
1 2 3 4 5 6 7 12 11 10 09 08

TABLE OF CONTENTS

CONSTRUCTION PERMIT

ACKNOWLEDGMENTS

The examples and stories in this book are inspired by real events that have occurred over the course of my career as a prosecutor or that I have heard about from the many inspectors I have worked with over those years. I would like to thank the people who have helped make this book possible, especially Robert McGinnis, the Building Commissioner of the Village of Hinsdale, Illinois, and Eric Alwin, the Building Official of the Village of Woodridge, Illinois. They were very generous with their time and related many examples of problems they had encountered dealing with contractors and homeowners. They were always available to review my manuscript and give me ideas for the content and structure of the book. Their technical expertise has been invaluable.

Kelly Anbach, a code inspector with the Village of Hinsdale, used her time and talent to take photographs for the book. I appreciate her generosity and value her friendship. Roger Axel, the Building Official for the City of New Hope, Minnesota, shared many photographs he had collected with me as did Kathy Hejnicki, a hardworking inspector with the Village of Woodridge, Illinois. Britt Pease, the president of the Minnesota Building Permit Technicians Association, enlisted members of that organization

to share their stories with me including Jackie Freppert of Arden Hills, Minnesota. Karyn Byrne, my friend and one of the best property maintenance inspectors I know let me use her name in some of the examples. Tim McElroy, a fire prevention inspector with the Hinsdale Fire Department, agreed to pose for photographs along with Robert McGinnis. Trevor Bishop, a fire prevention inspector with the Village of Bensenville, shared many useful examples of situations that went terribly wrong.

I must also recognize the contributions of Stacey Crockatt, an inspector with the Village of Lisle, Illinois, John Fincham, the assistant community development director with the City of West Chicago, Illinois, Tim Ryan, the deputy building commissioner of the Village of Hinsdale, and Scott Coren, the assistant to the city administrator in Darien, Illinois. I would like to thank my colleagues at the International Code Council and all of the building officials and inspectors who have taken my building code classes over the years. By sharing their experiences with me, I have gained the knowledge I needed to write this book. I wish to especially thank my administrative assistant, Genevieve Jones, who worked on the first draft of the manuscript.

Lastly, I wish to thank my husband, Alan, for taking the time to read the entire manuscript and giving me the benefit of his wisdom. There's nobody whose opinion I value more.

INTRODUCTION

The last thing anyone planning to build or renovate a home or other structure wants to think about during the planning process is local government regulation. If you are a contractor, your focus is on getting the job, obtaining plans, and lining up the subcontractors to do the work. If you are hiring someone to do the work for you, your energy is focused on obtaining estimates, finding the best builder for the job, and getting financing for the project. Nobody wants to think about building codes, permits, inspections, or certificates of occupancy.

Unfortunately, ignorance is not bliss when it involves ignoring building codes and other laws of the local jurisdiction. The consequences of ignoring local regulations may include fines, a delay in completing the project, hazardous conditions on the property, an inability to use the building, and difficulties when it comes time to sell the property.

Experienced builders understand the necessity of complying with the many laws that govern the building process. Unfortunately, too many people believe that if they just ignore the laws requiring permits and inspections, nothing will happen. They do so at their peril.

Figure Intro-1 Even seemingly minor changes to the structures on your property can require a permit; failure to acquire one is both illegal and can create unsafe conditions.

In the Field

"I once dealt with a man who thought he'd save money by not submitting plans or a permit application for an addition he wanted to build. He went ahead and excavated the land next to his home but didn't shore up the existing foundation. As a result, his first floor began to cave in and the foundation wall buckled. He ended up spending $30,000 just to make his existing structure safe again." —*Scott, Building Inspector* ■

This book will guide you through all the steps you need to follow when dealing with the legal requirements of local jurisdictions. It will help you understand why ignoring the law is foolish and against your own best interests. Though laws vary from town to town and state to state, the basic requirements of a building code are always the same. These include the necessity of submitting plans for the project, getting a permit, obtaining the required inspections, and securing a certificate of occupancy.

In Chapter 1 you will learn what building codes are, why they exist, and the scope of the various codes to get an understanding as to what type of code applies to your project. This is crucial information you need to know before you spend money on property and materials.

Chapter 2 deals with various factors you need to think about before embarking on a project. It will explain certain property law concepts such as boundary lines, required setbacks, covenants that run with the land, and easements. You will learn how the zoning code and stormwater management regulations could have a major impact on your endeavor. Local zoning laws could prevent you from using your building even if it meets the building and fire codes. Zoning laws dictate what the property can be used for and are important for both residential and commercial property. After reading this chapter, you will be able to research these issues so you are never in the predicament in which purchased property can't be used for its intended purpose.

In the Field

"I just wish people would call us before they buy property to renovate for a business. They think they'll just be able to do the renovations, move in, and open up their business. I had a guy with a sewer business who bought a property with a house on it. He wanted to use it for his office and to store his trucks outdoors. The problem was that the house didn't meet the current building or fire code requirements for a business, and the kind of outdoor storage he wanted wasn't allowed in that zoning district. The property also required thousands of dollars worth of landscaping and parking lot upgrades. I felt sorry for the guy, but I had to enforce the code. The property's up for sale and vacant and the buyer still doesn't have a place for his business."
—*Trevor, Building Inspector* ■

In Chapter 3 you will discover when you need a permit and when you don't. Repairs often don't require a permit but many other activities do, especially anything that affects the structural integrity of a building. A permit for new construction is required in any jurisdiction with a building code. Most building officials encourage people to contact them for information.

In the Field

"All people have to do is call. They're not bothering us even if it turns out they don't need a permit. We're happy to help them. We just want to make sure they're safe." —*Tim, Building Official* ■

Chapter 4 describes the permit application process in great detail. You will see what a sample permit application looks like so you can gather the necessary information for your application. Chapter 5 describes the types of documents you need to submit to the building department when you apply for a permit, and contains information on what you need to do if you want to change the plans. Chapter 6 explains the inspection process and why it's important for your safety. While many people who think they know what they're doing view inspections as a nuisance, inspections are critical for the protection of the public.

Figure Intro–2 Wiring should be done by an electrician familiar with the code. Improper installation can create dangerous living situations.

In the Field

"I went to do an inspection where a guy was installing a ceiling fan. When I did the inspection I found that he had used speaker wire instead of the proper electrical wire to hook it up to the electrical system. While the ceiling fan worked, eventually the wire would have overheated and burned his house down. Unbelievable." —*Rob, Building Inspector* ■

Chapter 7 explains the power that the building official has to issue a stop work order for unsafe work that could keep your project from getting finished in a timely fashion. Chapter 8 describes the ins and outs of getting a certificate of occupancy, which is something you will need before you can use or occupy the structure. Failure to obtain one is also one of the most common reasons homeowners end up in court.

In the Field

"We played cat and mouse with this family who had moved into their new house without a certificate of occupancy. We'd see their children catching the school bus and the lady of the house coming out to say goodbye to them in her bathrobe. When I tried to speak with the owner about performing the final inspection so they could get the certificate of occupancy, he insisted no one was living in the house and only workers were inside. It was crazy." —*Don, Building Inspector* ■

In Chapter 9 you will learn the unhappy consequences of violating the building code and strategies for dealing with a notice of violation. You'll be guided through the procedure you will face if you get cited for a violation and have to go to court. Ignoring a notice of violation can make things more difficult in the long run.

In the Field

"It was my own fault. As a contractor, I was trying to avoid permit fees when I did the build-out for a new business inside an office building. I got caught, had to take time off to go to court, paid double for the permits, and paid a $500 fine in court. Now the village inspector watches all of my projects closely and my professional reputation is damaged." —*Nick, Contractor* ■

Chapter 10 describes the appeal procedure to follow if you believe the building official made a mistake about a building code interpretation. In Chapter 11 you will learn about the procedure involved if you have to attend an administrative or court hearing because of a notice of violation. You will find out about strategies you can use when you engage in plea bargaining with the prosecutor.

In Chapter 12 you will gain insight into the pet peeves of building inspectors so you can avoid those traps and get your job done on time. Finally, Chapter 13 discusses what to do if you are unfortunate enough to be faced with corrupt, incompetent, or arrogant building officials and inspectors.

By paying attention to the building and fire code requirements in effect where your project is planned, you will save money, time, and aggravation. This book will be your guide to smooth sailing through city hall.

CHAPTER 1

BUILDING CODES

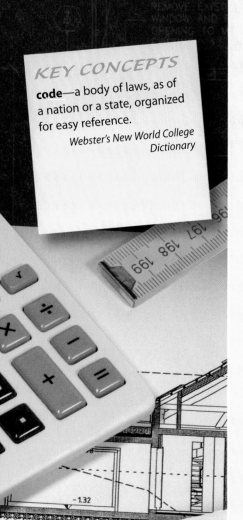

INTRODUCTION

Laws come into being when the legislative body of the federal, state, or local government passes a bill or an ordinance and it is signed by the president of the United States, the governor of the state, or the mayor or president of the local jurisdiction. This new law then becomes effective for the nation, state, or city. The laws discussed in this book are state and local government laws that regulate how buildings are constructed (building and fire codes) and how land is used (zoning laws). Laws passed by local government, for example, a city, town, village, county, or township, are called codes of ordinances. Therefore, a building code contains the laws that everyone must obey when constructing a building or other structure.

WHAT IS A BUILDING CODE?

A building code is a body of laws that governs how buildings are erected, constructed, altered, enlarged, repaired, used and occupied, maintained, or demolished. They are adopted by almost every type of local jurisdiction in order to ensure public safety. They may include a code for residential and commercial buildings, for fire prevention, and for property maintenance. The following sections will discuss these concepts in detail.

CONSTRUC

Purpose Of A Building Code

Most cities, towns, counties, and villages have building codes. The purpose of a building code is to make sure that the structures and the systems within them are safe and suitable for the purpose for which they are constructed by setting minimum standards to protect public safety, health, and the general welfare of the town's residents. Consider the stated purpose of the 2006 International Residential Code published by the International Code Council (ICC):

> **R101.3 Purpose.** The purpose of this code is to provide minimum standards to safeguard the public safety, health, and general welfare through affordability, structural strength, means of egress facilities, stability, sanitation, light and ventilation, energy conservation, and safety to life and property from fire and other hazards attributable to the built environment.

A building code describes all of the duties and obligations people have when they construct or alter structures or buildings. The main purpose of a building code is to protect the public and building occupants from dangerous conditions. Building codes came into existence because people have died and property has been destroyed because of unsafe buildings. This has been going on for as long as people have been engaged in construction. Even the Assyrians and the Romans had laws on constructing buildings. In the 1600s, London, England, adopted strict building code laws after great fires destroyed large areas of the city that had been built out of wood.

Figure 1–1   Construction laws have existed throughout history; even the ancient Romans regulated construction.

Different Kinds Of Codes

There are a number of codes that most cities adopt, including codes for commercial and residential construction, fire prevention, property maintenance, zoning, plumbing, electrical and mechanical systems, and stormwater management. These codes are developed over time by national organizations whose members are knowledgeable about the subject matter. Building codes are changed every few years as new techniques and materials are developed.

How The Codes Are Written

In modern times, organizations such as the International Code Council (ICC) and the National Fire Prevention Agency (NFPA) have developed a series of model codes covering commercial and residential construction so that each state or local government does not have to write a new code for itself. This saves the town a lot of money because it does not have to pay lawyers high legal fees to draft an original code. Other groups, such as the American National Standards Institute (ANSI), accredit standards developed by other organizations. These uniform standards are then referenced in the model codes so that buildings are safe throughout the nation.

How A Model Code Becomes A Law

Model codes promote uniform safety regulations. In some places, the state government decides which model code it wants to use as the building code for all of the communities in the state. The local cities, villages, and towns can usually adopt the state law as their local building code so that their inspectors can enforce the code. The local code cannot have weaker standards than the state code.

In other states, there is no state building code, but the state gives the power to regulate building construction to local governments. In those situations, the city, town, county, or village usually adopts a model code rather than incurring the cost of writing one for itself. It is adopted by reference, which means the entire content of the model code becomes part of the local jurisdiction's code of ordinances. The city can amend or change parts of the code if it wants to.

Model codes are promulgated by various organizations, such as the International Code Council (ICC) and the National Fire Prevention Association (NFPA). Some of the model codes that might be adopted include the following:

- International Building Code
- Uniform Building Code
- International Residential Code
- International Property Maintenance Code
- International Electrical Code
- National Electrical Code
- International Fuel Gas Code
- International Mechanical Code
- Uniform Mechanical Code
- International Plumbing Code
- Uniform Plumbing Code
- International Private Sewage Disposal Code
- International Fire Code
- National Fire Code
- NFPA Life Safety Code
- International Energy Conservation Code
- ADA Accessibility Guidelines
- ASME Boiler and Pressure Vessel Code

Why It Is Important To Know About Codes

When you begin a building project, it is important to know what codes have been adopted by the town in which you will be doing construction. This information is available at the city hall. You should investigate local codes before you spend time and money on plans, so you are following the correct law. If you hire a design professional, she is supposed to do this for you.

These model codes are updated every few years. Therefore, you must also find out what version of the building code has been adopted by the local jurisdiction. The International Code Council (ICC) holds code development hearings to update its codes every three years. In addition, states and local jurisdictions can amend any part of a model code. Therefore, you need to know what portions of the codes have been changed.

WHEN DOES THE BUILDING CODE APPLY?

Building codes cover a wide variety of activities. Before beginning a building project, you must find out if there is a code that applies to the work you want to perform. If you are in a state or town that uses the International Codes developed by the ICC, there are different laws that cover commercial and residential buildings. This section will discuss how to determine whether your project is covered by the requirements of the local building code. Ignoring the regulations of the building code can lead to terrible problems down the road.

In the Field

"One of the dumbest things I ever saw a person do without a permit was install a toilet. The person just ran the piping through the wall to the outside and never connected it to the sewer system. You can imagine how awful that was. If we didn't have a building code, people could do whatever they wanted, no matter how ridiculous." —*Kyle, Inspector* ■

Figure 1–2 Building codes also regulate plumbing installation.

Scope Of The Building Code

The following table is a comparison of the scope of the International Building Code versus the International Residential Code. It tells

TABLE 1-1   Scope of the International Building Code Versus International Residential Code.

SCOPE OF THE INTERNATIONAL BUILDING CODE—IBC 101.2	SCOPE OF THE INTERNATIONAL RESIDENTIAL CODE—IRC R101.2
The IBC code applies:	**The IRC code applies:**
To every building or structure or any appurtenances connected or attached to the buildings or structures when there is:	To every detached one- and two-family dwellings and townhouses not more than three stories above-grade height with a separate means of egress and their accessory structures when there is:
■ Construction	■ Construction
■ Alteration	■ Alteration
■ Movement	■ Movement
■ Enlargement	■ Enlargement
■ Replacement	■ Replacement
■ Repair	■ Repair
■ Equipment use	■ Equipment use
■ Occupancy	■ Use and occupancy
■ Location	■ Location
■ Maintenance	■ Removal or
■ Removal or	■ Demolition
■ Demolition	

Does The Code Cover The Construction Project?

JUST ✔ CHECKING

Are any of the following activities included in the construction project?

- ☐ Construction
- ☐ Alteration
- ☐ Movement
- ☐ Enlargement
- ☐ Replacement
- ☐ Repair

- ☐ Equipment use
- ☐ Occupancy
- ☐ Location
- ☐ Maintenance
- ☐ Removal or
- ☐ Demolition

---

you what type of activity is regulated and which code applies to the type of building involved. If the activity you want to perform is in this list, the building code applies to your project.

How To Determine Which Code Covers The Construction Project

It is important to determine which building code covers your project. Different codes cover different kinds of jobs. Typically, there may be one code that regulates commercial construction and large residential projects and another one that governs small residential projects. For example, the International Residential Code published by the International Code Council applies to:

- Detached one- and two-family dwellings and townhouses not more than three stories above-grade in height with a separate means of egress, and their accessory structures
- Existing buildings undergoing repair, alterations, or additions and change of occupancy

The International Building Code does not apply to these types of buildings but applies to all other types, such as commercial buildings and multi-family dwellings such as apartment buildings.

Figure 1–3   The International Building Code applies to commercial buildings.

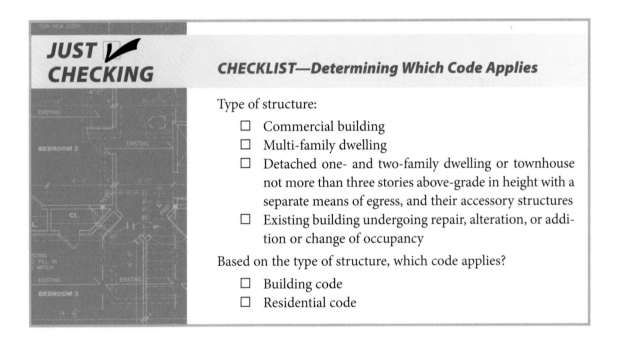

JUST ✔
CHECKING

CHECKLIST—Determining Which Code Applies

Type of structure:

- ☐ Commercial building
- ☐ Multi-family dwelling
- ☐ Detached one- and two-family dwelling or townhouse not more than three stories above-grade in height with a separate means of egress, and their accessory structures
- ☐ Existing building undergoing repair, alteration, or addition or change of occupancy

Based on the type of structure, which code applies?

- ☐ Building code
- ☐ Residential code

Types Of Permits Covered By The Code

Other codes are important in the construction process, including the local fire, plumbing, electrical, and mechanical codes. Once you have determined which code applies, you must then figure out if you need a permit to do the work. The building code that applies will describe what type of work you need a permit for and when you don't need a permit. Typically, a building code governs construction, electrical, mechanical, and plumbing work. There may be special rules that deal with fuel gas, sewage disposal, stormwater management, energy conservation, accessibility for disabled persons, and fire prevention.

Figure 1–4 The International Building Code or International Residential Code applies to condominiums depending on the size of the building.

Figure 1–5 The International Residential Code applies to single family dwellings.

Does My Project Include:

JUST ✔ CHECKING

- ☐ Building
- ☐ Electrical
- ☐ Fuel gas
- ☐ Mechanical
- ☐ Plumbing
- ☐ Private sewage disposal
- ☐ Fire prevention
- ☐ Energy conservation
- ☐ Changing the grade of property

How To Determine Whether The Fire Code Applies

Fire codes are very important in almost any kind of construction. The requirements range from smoke detectors to fire sprinkler and alarm systems tied directly to the fire department. In many jurisdictions, the building department makes sure that the fire code requirements are followed.

Local jurisdictions either have their own fire department or are part of a group of other jurisdictions covered by a fire district. The fire code you must follow is the one adopted by the state and the local jurisdiction, if there is a fire department, or by the fire district. You must check with the authority having jurisdiction over fire protection to find out which code you must follow. Sometimes both the NFPA fire codes and the International Fire Code are adopted and you must follow all of them.

Whenever you are doing a project, you must also find out whether the work is within the scope of the local fire code. The following table describes when the fire code applies.

Figure 1–6 An electrician must know the scope of the local fire code when installing a fire alarm system.

TABLE 1-2   Applicability of Construction and Design Provision of the International Fire Code.

FIRE CODE APPLIES TO:	IFC SECTION NUMBER
Structures, facilities, and conditions arising after the adoption of the code	102.1(1)
Existing structures, facilities, and conditions not legally in existence at the time of the adoption of the code	102.1(2)
Existing structures, facilities, and conditions when identified in specific sections of the code	102.1(3)
Existing structures, facilities, and conditions which, in the opinion of the fire code official, constitute a distinct hazard to life or property	102.1(4)

As you can see, the fire code applies to a wide range of buildings. You should check with the local fire department or district to find out if your structure is covered.

The following table explores the scope of the fire code.

TABLE 1-3   Scope of the International Fire Code.

THE CODE APPLIES TO STRUCTURES, PROCESSES, PREMISES, AND SAFEGUARDS REGARDING:	IFC SECTION NUMBER
The hazard of fire and explosion arising from the storage, handling, or use of structures, materials, or devices	101.2(1)
Conditions hazardous to life, property, or public welfare in the occupancy of structures or premises	101.2(2)
Fire hazards in the structure or on the premises from occupancy or operation	101.2(3)
Matters related to the construction, extension, repair, alteration, or removal of fire suppression or alarm systems	101.2(4)

If your proposed building includes any of these conditions, you must follow not only the building code, but the fire code too.

AVOIDING TROUBLE

- Be familiar with the codes of the jurisdiction in which your building project is located.
- In you lack the necessary skill to interpret the codes, hire someone who does.
- Make sure you know what version of the code is in effect.
- Find out what codes may apply to your project.
- Find out if the fire code applies to the project.

CONCLUSION

Building projects are governed by a number of different building and fire codes. The codes ensure that public welfare is protected and that structures meet minimum safety standards. Finding out which codes apply to your project is vital. It is quite apparent that doing almost any type of construction will mean that you should call the local building department and find out what you need to do to comply with the local codes. If you think you may be doing a small project that is exempt from the permit requirement, you'll find out more information in Chapter 3. But, it never hurts to call your local building department.

CHAPTER 2

MATTERS TO CONSIDER BEFORE BUILDING

INTRODUCTION

This chapter will explore important issues you need to take into consideration before undertaking your building project. Three areas are of special concern: property law issues, zoning, and stormwater management. Each will be discussed in detail in the following sections.

PROPERTY LAW ISSUES

There are a number of property law concepts that every homeowner or builder must understand in order to not run afoul of the law. Many people think that because they own land, they can do anything they want upon it. Unfortunately, this is not true because of the many different types of restrictions that are placed on property by various statutes and ordinances. These concepts go back hundreds of years and are still relevant today. Therefore, the wise homeowner or contractor will keep these in mind when planning a project. This section will discuss important issues to consider.

Property Lot Boundaries

Many people believe the mistaken idea that their property lot extends all the way to the curb of the roadway. You can find out the exact dimensions of your lot by looking at a survey. Usually you obtain a survey at the time that you purchase the property from the seller. The survey is drawn by a licensed

KEY CONCEPTS

survey—to determine and delineate the form, extent, and position (as of a tract of land) by taking linear and angular measurements and by applying the principles of geometry and trigonometry.

Merriam-Webster's Online Dictionary

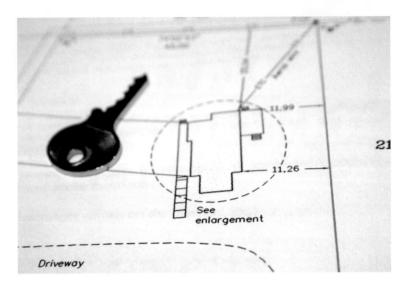

Figure 2–1 Example of a property survey.

surveyor who is trained in finding out exactly where the property lines begin and end. A survey will show where the structures are on the property as well as whether there are any easements or setback lines. These concepts will be discussed a little later in this section.

The important point is that when you begin a building project that requires a survey, you should hire a licensed surveyor so that you can find out exactly where your property begins and ends and where the important imaginary lines exist on the property. Most property lots end at the sidewalk. The street of the local jurisdiction does not usually end at the curb but extends a few feet further on both sides of the roadway, typically referred to as a public right of way. While you may have cut the grass for 10 years on the parkway in front of your residence, it is

In the Field

We are very strict in our village on construction problems involving the village parkway. We don't allow people to store their construction materials upon it and can get a $500 ticket if they do. We require that they protect trees on the village parkway by putting up a fence and a bond so that if a tree is damaged in any way, the permit holder is responsible for paying for a new one. People just have difficulty understanding that they don't own that piece of the land but if they would just look at their survey they would see that their front yard lot line does not extend to that particular area. —*Rob, Building Official* ■

Figure 2–2   Most property lots end at the sidewalk. Parkway trees are usually under local jurisdiction; some towns institute measures in order to protect them.

actually property owned by the local jurisdiction. Therefore, there are restrictions that apply to the use of that property. For example, some towns institute measures to protect their parkway tree.

It is very important to know exactly where your lot line is by having a survey done so that if you erect something, such as a fence, you don't erect it on the neighbor's property. An encroachment could create a problem for you or the neighbor when it is time to sell either property.

Figure 2–3   Check your land survey before building a yard fence to ensure that it does not encroach upon your neighbor's property.

How To Determine The Boundaries Of A Lot

When you buy property you obtain a deed to the property, and the deed contains information regarding the legal description of your property. Property is normally broken down into large areas that are then further broken down into blocks, which are then further broken down into lots. The local county keeps books filled with plats of surveys that show drawings of entire neighborhoods and their lots.

Legal descriptions are so specific that any surveyor can determine exactly where property lines begin and end. A legal description may read something like:

> Lot 2 in Block 5 in Schiller's Addition to Hinsdale, a subdivision in the Southeast quarter and the Southwest quarter of Section 8, Township 38 North, Range 11, East of the Third Principal Meridian, in DuPage County, Illinois.

You will find the legal description for any property on the deed that was issued when the lot was purchased and in the title report issued at that time. A surveyor uses this information when doing a survey of the property. Typically, there are pipes located at the four corners of the property the surveyor can use to stake out the exact lot lines of the property.

Figure 2–4   A surveyor conducting a survey on a residential construction site.

Required Setbacks

A setback is the specific number of feet set forth in the local zoning code in which no building may be placed. There is usually a setback on all sides of the property so that there is no building that is too close to the neighboring property or the street. Every setback differs from town to town depending on the zoning code. You must find out what the setback requirements are for your particular piece of property if you are planning to build a structure upon it.

In the Field

"Mr. and Mrs. M built a large home but then deviated from the plan by placing the air-conditioning units in the side-yard setback area. An inspector caught this during one of the routine inspections and told them that they would have to remove the concrete pad for the air-conditioning units because it couldn't be within the six feet side-yard setback. Mr. M refused to do so. We had to take them to court and have the court order them to remove the concrete pad. They were upset because it looked better where they wanted to put it and it was quieter for their family. Unfortunately, they were so close to the next-door neighbor that those neighbors complained. Ultimately, it cost them probably close to one thousand dollars between the fines and the legal fees for something that they should have never done in the first place." —*Scott, Building Inspector* ∎

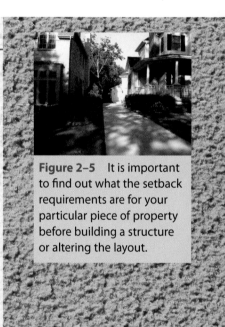

Figure 2–5   It is important to find out what the setback requirements are for your particular piece of property before building a structure or altering the layout.

Covenants

A covenant is an agreement or promise between two or more parties to either do something or refrain from doing something. Covenants can be affirmative, negative, or restrictive. Covenants are usually recorded at the time that the property is sold and run with the land. That means that every subsequent purchaser must obey the restriction in the covenants. Some covenants restrict how the property can be used. For example, a deed may contain a restriction that a gas station can never be built on the property. If you are planning to build a gas station and buy this type of property,

KEY CONCEPTS

covenant—an agreement or promise between two or more parties. A covenant that runs with the land goes with the land, cannot be separated from it or transferred without it.

Black's Law Dictionary

you could end up being sued by the prior owner to return the land to him because you violated the covenant.

Other covenants are imposed by groups such as homeowners' associations. In some homeowners' associations, there are numerous restrictions on whether you can have a fence, what color you can paint your house, and whether you can have any accessory structures. The important thing to realize is that many of these restrictions are not illegal under the building code or the zoning code in the local jurisdiction. Therefore, it is possible that you could obtain a permit to build a shed or a fence and obtain the necessary permits from the local jurisdiction yet be sued by the homeowners' association and be forced to remove the structure. It is not the responsibility of the local jurisdiction to inform you of any restrictive covenants that run with your land. If you are given any information, it is merely as a courtesy and you cannot hold the local jurisdiction responsible if you violate a restrictive covenant.

When you buy your property, the deed indicates that along with title to the property you are also taking it subject to any restrictions, covenants, and assessments of record. Thus, when you purchase property it is important to obtain a title search from a title company that will research the property history for these kinds of restrictions.

Title companies research all of the documents in the Recorder of Deeds office that have anything to do with your property, such as deeds, mortgages, liens, or other restrictions such as a covenant. Prior to the real estate closing, the title company prepares a report and has copies for both the buyer and the seller. The buyer pays the title company for insurance to guarantee that if the company fails to find that an encumbrance or restriction exists, you have a way to recover your damages if you did not get clear title to the property or there was a restriction of record that was not uncovered.

Easements

An easement entitles a person who does not own the land to use a portion of the land possessed by the owner. Typical easements that are found on property include utility easements so that the electric or water company can come on the land to fix problems as they arise. For example, the electric company can come on your property and trim branches that may interfere with electric wires. This is why you cannot put a permanent structure on an easement.

KEY CONCEPTS

easement—an interest in land owned by another person, consisting in the right to use or control the land, or an area above or below it, for a specific limited purpose (such as to cross it for access to a public road).

Black's Law Dictionary

Figure 2–6   Utility easements ensure that the electric or water company can come on the land to fix problems as they arise without being hampered by privately owned structures.

If you are planning to put in a pool or build a garage, you need to make sure that you put it where there are no easements so you cannot be forced to remove it in the future.

Another type of easement allows someone to go across your property in order to get to his property. This often happens where the owner of two or more adjacent lots sells part of the property and grants the purchaser easements needed to use the property. If there is this type of a road across the property where you plan to build, you cannot block this easement that runs with your land or you may end up being sued by the person who has the right to use the property because of this prior agreement in the deed.

ZONING ISSUES

Your building plans may be flawless and the location may be perfect, but your project may be doomed because of the zoning ordinances of the municipality or county. What are these regulations that make life so difficult for individuals? If you own the land, why can't you do what you want? Researching zoning issues is essential before buying a piece of property or starting a project.

> *KEY CONCEPTS*
>
> **zoning**—the division of a city by legislative regulation into districts and the prescription and application in each district of regulations having to do with structural and architectural designs of buildings and of regulations prescribing use to which buildings within designated districts may be put.
>
> *Black's Law Dictionary*

In the previous example, Mr. F lost everything because neither he nor his lawyer checked the zoning laws of the local jurisdiction before he bought the property. Zoning laws give municipalities and counties control over how land is divided up and used. The local jurisdiction can regulate land use in an orderly manner. The underlying purpose is to protect the public welfare and to maximize the value of the land. You don't want to build a home only to have a garbage dump appear next door to you a couple of years after you move into the neighborhood. Imagine how difficult it would be to sell your residence if that happened.

How Zoning Laws Work

Zoning laws divide up the local jurisdiction into districts. In each of these districts, certain types of buildings and uses of the land are allowed. This prevents someone from building an office park in the middle of a single-family residential area. It also prevents a person from running an automobile repair shop out of his family home's garage. As a general rule, if the use is not listed, it is not allowed.

Municipalities and counties are often divided up into the following different types of districts:

- Single-family residential
- Multi-family residential
- Office or commercial
- Retail
- Manufacturing or industrial

KEY CONCEPTS

use—the purpose or activity for which the land or building thereon is designed, arranged or intended or for which it is occupied or maintained.
Village of Woodridge Code

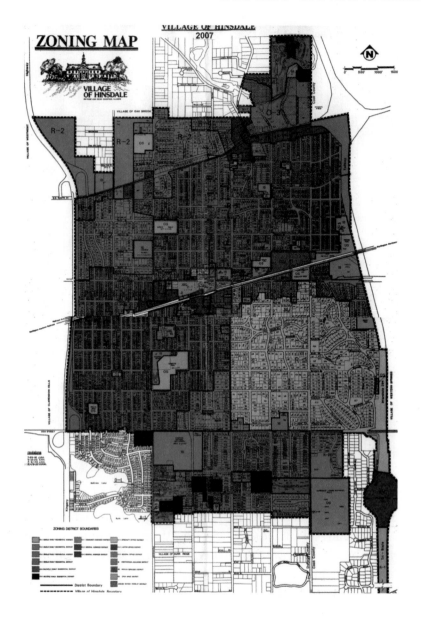

Figure 2–7 A zoning map shows how the municipality or county is divided up into zoning districts.

Within a zoning classification, there are usually smaller districts within the larger district regulating density and uses. They are usually differentiated by assigning a number to the type of district they are in, such as R-1, R-2, R-3, and so on for residential areas and I-1, I-2, I-3, and so on for industrial districts. Certain uses that might be allowable in an R-2 zoning district might not be allowable in an R-1 district. For example, in some types of residential districts, schools, parks, and churches are

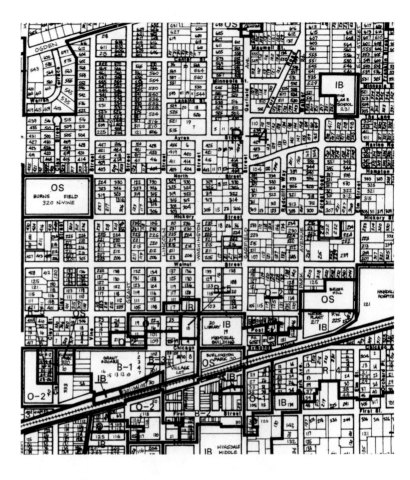

Figure 2–8 In addition to consulting a zoning map, speaking with someone who has authority in the zoning department is always a good idea when trying to determine if a property is zoned for your intended purpose.

allowed in addition to homes, but this regulation might not be true for all residential district classifications. An animal hospital might be allowed in a particular type of business district, but not in all of them.

A typical zoning code will list the types of uses that are allowed. For example, here are the listed uses for this R-1 residential district:

9-5A-1: USES PERMITTED:

Accessory uses.

Churches as ancillary uses to schools (elementary, junior high and senior high) only.

Churches, convents, rectories, parsonages; provided that such uses shall only be permitted along collector streets.

Parks and other recreational areas owned by the village or other governmental unit or private homeowners' association.

Public service and utility uses.

Single-family detached dwellings.

State of Illinois licensed group home.

Village of Woodridge Zoning Code

Every local jurisdiction should have a zoning map available to the public that shows how the municipality or county is divided up into zoning districts. You should look at the location you are concerned about and see how it is classified. The map will have a key, as shown in the accompanying figure, that will show you on the map how the area is zoned.

You then need to read the zoning ordinance for that classification to see what uses are allowed. You should also speak with a person having authority in the zoning department if it is not clear whether you can do what you intend in that particular district. Remember, speaking to the receptionist doesn't count.

Don't Attorneys Take Care Of These Issues?

Why doesn't your attorney find out about these problems before you buy the property? The reason is that most attorneys know very little about zoning laws. The average real estate attorney wants to make sure you get clear title to the property you are purchasing and that there are no outstanding liens or encumbrances. The attorney will not usually ask you what you intend to do with the property or investigate the structure's legal status. If the seller misrepresents what the property can be used for, you may have a potential lawsuit for fraud, but that doesn't help you run your business while the lawsuit is pending. You can't rely on your attorney to find out about these issues. You should ask the attorney to verify that the property can be used for the purpose you intend. Ultimately, someone must call the proper person in the local jurisdiction to resolve the legal issues. Get the zoning official's finding in writing if possible.

The following form can be used to help you answer these questions.

REAL ESTATE WORKSHEET—ZONING

Zoning issues:

What is the intended use of the property?_____

What is the zoning district for location of property?_____

Is the use allowable? Yes _____ Yes, with special-use permit _____ No _____

Is it likely the special-use permit will be granted? Yes _____ No _____

Is getting a special-use permit a condition of the sale? Yes _____ No _____

Will the fire protection system have to be upgraded for the intended use?
Yes _____ No _____

Are there any legal-nonconforming-use or structure issues with the property?
Yes _____ No _____

If yes, will the use of the property by the buyer extinguish the legal nonconformity?
Yes _____ No _____

If yes, will any alterations to the property extinguish the legal nonconformity?
Yes _____ No _____

Does the buyer need a certificate of zoning for the property before it can be occupied?
Yes _____ No _____

Home occupation:

If it is residential property, will there be a home occupation?
Yes _____ No _____

If yes, nature of the business: _____

Number of employees:_____ Number related to the buyer? _____

Is business totally conducted within the residence? Yes _____ No _____

Outdoor storage of equipment or vehicles: Yes _____ No _____

Is the use allowable? Yes _____ No _____

Illegal conversions:

If the residence is for multi-family use, is it a legal structure or use? Yes _____ No _____

Over-occupancy:

Is the residence large enough for the number of people who intend to live in it?
Yes _____ No _____

Figure 2–9 Real Estate Worksheet.

Special Or Conditional Uses

Some uses are special or conditional uses and may be allowed in a district only with the consent of the zoning board. There may be special conditions placed on the use of the property if this use is granted. It is often very expensive to apply for a special-use or conditional-use permit. You should never buy property believing you will be granted a special-use permit unless the purchase is contingent on obtaining the special use. You would be wise to hire an attorney who specializes in this type of law. It is a complicated and expensive process to undertake, and there is no guarantee you will get one if you apply. An attorney can give you advice on whether it is even worthwhile to pursue.

Here is an example of special uses in a business district:

9-6A-2: SPECIAL USES:

Automotive parts and accessory stores, not including automotive machine shops.

Daycare centers and preschools.

Drive-in and drive-through facilities, accessory.

Electronic message board signs, provided that they serve as shopping center identification signs and that they meet all applicable regulations set forth in Chapter 11 of this title.

Fast food restaurants.

Planned unit developments.

Public or private facilities such as libraries, hospitals, institutions, government buildings and other similar uses.

Village of Woodridge Code

No one can operate any of these businesses without applying for a special-use permit and participating in a special-use hearing. At the hearing, residents of the municipality may appear and object to the type of business you wish to pursue for a number of reasons, including increased density, traffic, and safety issues.

Accessory Uses

Accessory uses are allowed in all districts. For example, a garage can be built for a single-family home. A retail store may have a parking lot. You don't want to assume something is an accessory use. There are certain standards that accessory uses must meet. If you are uncertain about an accessory use, call the zoning inspector for the local jurisdiction.

KEY CONCEPTS

accessory use—an accessory use is a structure and/or use which:

A. Is subordinate to and serves a principal structure or use;

B. Is subordinate in area, extent, intensity and/or purpose to the principal structure or use served;

C. Contributes to the comfort, convenience or necessity of the occupants of, or of the business or industry located in or on the principal structure and/or use served; and

D. Is located on the same zoning lot as the principal structure and/or use served.

Village of Woodridge Code

Figure 2–10 A garage is an example of accessory use.

Home Occupations

Like any other business, home occupations are regulated by the zoning code. Businesses that can be conducted totally within the residence are usually allowed. Those that have outdoor storage or attract a lot of traffic may not be allowed. Do not assume you can use a residence or garage to conduct your home-based business. There are consequences to violating the zoning code.

In the Field

"We once had a guy who was operating a full-service auto repair shop in his large garage. Cars would be dropped off in the morning and picked up at night. The neighbors were going crazy. We finally were able to catch him and shut him down after we had a police officer from our town get an oil change in the garage. He saw hydraulic lifts and all kinds of tools inside the garage when he dropped off his car." —*Richard, Zoning Inspector* ■

Here is an example of a local ordinance with standards regulating home occupations:

5-22-4: STANDARDS:

(A) The operator(s) of the home occupation shall make the dwelling unit within which the home occupation is conducted his/her legal and primary place of residence.

(B) One person at a time who is not a resident of the dwelling may be permitted to assist in the home occupation.

(C) The home occupation shall not regularly or frequently attract more than four (4) people simultaneously to the premises for reasons related to the home occupation, with the noted exception only of state licensed daycare homes and part day childcare facilities/ services as described in subsection (J) of this section.

(D) The home occupation shall be subordinate and incidental to the use of the dwelling unit as a residence. Among other considerations, a home occupation will be considered to be subordinate and incidental if it occupies thirty five percent (35%) or less of the gross square footage of all floors of the house, including the basement. Such gross square footage shall not include the area of a garage, accessory or ancillary building. Further, the home occupation may be conducted anywhere within the house, including the basement, and within a garage located upon the same lot as a residence from which a home occupation is conducted in any ancillary building upon the property, as defined in section 5-1-7 of this title. For example, if the total area of a house, including all floors, is equal to one thousand (1,000) square feet, the home occupation may occupy only three hundred fifty (350) square feet in order to be considered subordinate and incidental. Whether the home occupation itself is conducted within the house or in a garage, three hundred fifty (350) square feet would be the maximum area that could be devoted to the home occupation.

(E) No home occupation shall cause an increase in traffic or parking congestion in excess of those which normally occur in that residential neighborhood.

(F) Supplies and equipment shall be delivered by U.S. postal service, parcel delivery services, private passenger automobiles or vans, or by trucks that customarily make deliveries to residences that do not have home occupations, such as, but not limited to, those servicing furniture and appliance stores.

(G) There shall be no signs, advertising, display, or activity that will indicate from the exterior of any building that it is being used, in part, for any use other than dwelling. No storage or display of materials, goods, supplies or equipment related to the operation of a home occupation, including vehicles associated with the operation of a livery service operating from the residence, shall be visible from the outside of any building located on premises on which the home occupation is conducted. Further, the parking of vehicles within residential districts must be in compliance with all village ordinances, including those restrictions on heavy weight vehicles as set forth in subsection 5-13-1-5(A) of this title.

(H) All home occupations must comply with all restrictions set forth in this code regarding noise, odors, radiation, solvents, electronic interference or any other consequence, result or byproduct of the home occupation. Further, no home occupation may generate noises, odors or any other byproduct in excess of the level normally occurring in that residential neighborhood.

(I) Any home occupation required to be licensed by the state or any other governmental body shall not be permitted until and unless a current license from such body is obtained.

(J) Babysitting services, daycare homes and part day childcare facilities/services, as defined by Illinois Compiled Statutes, shall comply with the standards of the state, take a maximum number of eight (8) children, which includes the natural or adopted children of the person operating the babysitting service or daycare home under the age of fourteen (14). Said services shall be allowed to operate within that home and have limited supervised outdoor activity within the compliance of the state statute on childcare. This subsection shall not be construed to permit a daycare center, childcare institution, daycare agency, group home or child welfare agency, as defined in the Illinois Compiled Statutes.

(K) No use shall require internal or external alterations or involve construction features or the use of electrical or mechanical equipment that would change the fire rating of the structure or the fire district in which the structure is located.

(L) The home occupation shall not require internal or external alterations or construction features not customary in residential areas.

(M) The home occupation shall not interfere with the reasonable use and enjoyment of neighboring residential properties and shall be located or conducted so that the average neighbor, under normal circumstances, would not be aware of its existence.

(N) No home occupation shall be operated in such a manner as to cause a nuisance or be in violation of any applicable statute, ordinance or regulation.

(O) The home occupation shall be conducted in a manner which would not cause the premises to differ from its residential character either by use of colors, materials, lighting or the emission of sounds, noises, or vibrations.

(P) The building or structure in which the home occupation is located shall be subject to the regulations of the zoning district in which it is located, except as provided herein.

Village of Lisle Zoning Code

In the Field

"We had a contractor in town who ran everything out of his home. He had a large lot and he didn't bother anybody. He was part of the founding families in town. The area began to change as younger, more wealthy people moved into the area. When one of them complained about his workers loading up their trucks at his home, I finally had to tell him it was time to move his business out of his home." —*Karyn, Zoning Inspector* ∎

Sometimes illegal home businesses have lasted for years in a particular area but then the demographics of the area change and the new neighbors are not nearly as tolerant as the old ones. The zoning officer investigates, and the owner of the business then needs to find a new location to conduct the business.

Nonconforming Property

A property may be used even if its use or the structure on it violates the current zoning code because it has been used for its current purpose for so long that it predates the adoption of the zoning code. As long as the use does not change and it was legal to begin with, the building may continue to be used as it has in the past. For example, a two-flat in a single-family residential zone may continue to be used as a two-flat if it was a legal use when the zoning code was adopted. A small grocery store whose use predates the adoption of the zoning code may continue to be operated in a residential district. However, once the use or the structure is changed, the legal nonconforming use is lost and the current zoning code must be followed. Ignoring this issue can have drastic consequences.

In the Field

"Mrs. C bought a small sandwich shop in a residential district. She decided business was so good that she would expand the building, but she didn't check with the building official about the effect this would have on her zoning classification. Her changes to the structure were so extensive that the city ruled that she lost her legal nonconforming status and she was forced to close the business." —*Kelly, Code Inspector* ∎

Sometimes a person buys property believing the use is legal but, in fact, it is an illegal nonconforming structure. For example, an owner converted a single-family residence in a single-family residential district into a building with three separate apartments 15 years ago. The local jurisdiction did not know about it because no one ever complained. The realtor listed the property as a three-flat and the buyer purchased it with the intent to rent two of the apartments so he could afford the mortgage. After the sale went through, the zoning official found out about the illegal use and forced the owner to reconvert the property to its legal state, a single-family residence. The buyer could not afford the mortgage without the renters and had to sell the property.

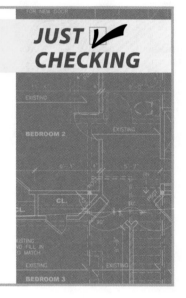

Zoning Issues

JUST CHECKING

Before you buy or rent property, you must ask yourself the following questions:

1. Is the property suitable for my intended use?
2. Can the structure be used for my intended use?
3. Will I have the right to occupy the property after buying it without making any changes?
4. Are there any legal nonconforming issues with the property?
5. Is the structure legal?
6. Will the local jurisdiction issue a certificate of zoning?

Exercise due diligence before spending a penny on plans for property by calling the zoning and/or community development departments in your local jurisdiction to make sure you can use the property for your intended purpose, especially if it is different from the surrounding properties.

STORMWATER MANAGEMENT

An overlooked pitfall for homeowners and contractors doing any kind of project is the impact the plans may have on the local environment, especially stormwater management issues. Ever increasing regulations regarding this issue can force you to change your plans or greatly increase the cost. Most people don't realize that if they change the grade of the property or if they are building in a flood plain, they need a special permit in addition to the regular building permit.

Benefits Of Stormwater Management

The purpose of stormwater management is to reduce the risk for damage to persons and property in the event of a flood. Engineers try to anticipate what might happen in the event of a 100-year flood and plan accordingly.

Figure 2–11 Wetland preservation is considered in the regulatory scheme of stormwater management.

In the Field

"We sent violation notices to a number of homeowners whose properties bordered a small creek. The homeowners had built gazebos and footbridges across the streams, all without building permits. What they didn't understand is that when the creek becomes a river in a year of unusually heavy rain, those objects become guided missiles when they are torn lose by the raging water. They become lethal weapons against anybody and anything in their path." —*Stacey, Code Inspector* ■

Wetland preservation is also part of this regulatory scheme, which has the benefit of reducing flood risk and preserving the environment for local vegetation and wildlife.

Who Needs A Special Permit?

If you don't know whether you need a grading or stormwater management permit, you must follow the same rule as for any kind of new construction. Read the code and ask your local building department. You shouldn't be able to pass a plan review if you lack the proper documentation, so ask the building official before you incur expenses. It is the safest course of conduct. The local jurisdiction often regulates these issues for the county and state, but it is possible that you may need to get a permit from the county as well as the local jurisdiction. The building official will be able to direct you to the right agency.

Why Build In A Flood Hazard Area?

Only you can decide whether the benefits outweigh the detriments. Houses are usually cheaper in a flood hazard area, but to protect yourself you will need extra insurance, assuming you can get it. And, above all, don't just meet minimum requirements. You can't assume that insurance will compensate you for all of your losses. No amount of money can compensate you for the emotional devastation that happens during a flood.

Grading Permits

Grading permits are needed so water flows and drains properly throughout the neighborhood. Disputes between neighbors are common if one person changes the grade of the land to the benefit of his property but creates a pond in the neighbor's backyard. Sometimes changing the grade does more than cause improper drainage. People can disturb wetlands and create terrible trouble for themselves.

In the Field

"I had one guy who didn't like the way water drained on his property, so he changed the grade of his property without a permit. What he didn't know was that part of the area he disturbed was a wetland. He was really stubborn and didn't want to fix the problem once we caught him. Between the topographical survey, the engineering study, legal fees and fines, he must have spent $10,000. One phone call to me and he could have avoided it all. People are their own worst enemies." —*Scott, Building Official* ∎

AVOIDING TROUBLE

Property Issues

- Whenever you are planning a construction project, obtain a survey by a licensed surveyor.
- If you are planning to purchase property, obtain a title report from a title company in order to find out what types of restrictions run with the property.
- Never buy property without knowing what the restrictions are on the property.
- Do not rely on the local jurisdiction to tell you about any restrictive covenants.
- Make sure you properly position any structure you are planning to build so that it does not encroach upon any easements, setbacks, or neighboring property.

Zoning

- Always check with the local jurisdiction before buying or renting property to make sure you can use the property for your intended purpose.
- Don't spend any money on building plans until you check the zoning code.
- Always investigate whether a structure is legal for its intended use with the local jurisdiction.
- If you will need a special-use permit, find out the total cost of applying for one so you can make an informed decision as to whether it's worth the cost, whether your request is reasonable, and whether there will be public opposition to your plans.

Stormwater Management

- Never perform any work that changes the flow of water over your property or the grade of the land without checking with your local building department to find out if you need a permit.
- If you need a stormwater management permit, apply for one.
- Don't begin work until your permit is approved.
- Follow the approved plans and don't deviate unless you get approval.

CONCLUSION

Having beautiful building plans is not sufficient for a construction project. You need to find out what restrictions of record, zoning, and stormwater management issues exist that may derail your plans. There is no sense in constructing a building you will not be able to use because the zoning ordinances do not allow the use you intend at the location you have selected.

To avoid surprises, before spending a lot of money on building plans, always check with the local building and zoning department to find out if your project is feasible and the land can be used for your intended purpose. Check the title report to make sure there are no private restrictions that would bar your use of the property. Also, do adequate research to make sure that the cost of your project will not be drastically increased by required stormwater management restrictions.

CHAPTER 3

PERMITS-EXEMPT WORK

INTRODUCTION

Not every type of work requires a permit. A permit is an official document or certificate issued by the authority having jurisdiction (e.g., a city, county, township, etc.) that gives permission or authorizes the applicant to conduct the specific activity described in the permit application in the manner approved by the building department. Permits cover a wide range of activities. Those activities have already been described in Chapter 1. This chapter will discuss the types of activities and repairs that do not need a permit.

EXCEPTIONS TO THE PERMIT REQUIREMENT

Some types of activities are exempt. The model International Building Code and the International Residential Code describe the kinds of activities that can be performed without a permit. These activities are described in the tables that follow. When you look at the column in each table marked "Type of Work," make sure you read the "Specific Requirements" that go along with it. For example, if an accessory structure used as a tool or storage shed exceeds a floor area of 120 square feet, the exemption no longer applies and you need to get a permit. The work is exempt-only if it complies with the specific requirements. This is important. You cannot just look at the first column and think the law does not apply to you.

39

Figure 3–1 An example of a three car garage for which a permit is needed.

Figure 3–2 An example of a small shed for which a permit is not needed.

TABLE 3-1 Work Exempt from Permit—International Residential Code.

TYPE OF WORK	SPECIFIC REQUIREMENTS
Accessory structures—one story	Used as tool and storage sheds, playhouses, and similar uses if floor area does not exceed 120 square feet (11.15 m²)
Cabinets or similar finish work	None
Carpeting or similar finish work	None
Cooling unit, portable	None
Countertops or similar finish work	None
Electrical repairs and maintenance	Minor work including replacement of lamps or connection of approved portable electrical equipment to approved permanently installed receptacles

TABLE 3-1   Work Exempt from Permit—International Residential Code.   *continued*

TYPE OF WORK	SPECIFIC REQUIREMENTS
Emergency replacement of equipment and emergency repairs	Must submit permit application within the next working business day (IRC R105.2.1)
Evaporative cooler, portable	None
Fences	Not over 6 feet (1,829 mm) high
Gas clothes-drying appliance	None
Gas cooking appliance	None
Gas replacement of parts	Does not alter approval of equipment or make it unsafe
Heating appliances, portable	None
Leaks in drains and water, soil, waste, or vent pipe: stopping of leaks in water, soil, waste, or vent pipe	If any concealed trap, drain pipe, or water, soil, waste, or vent pipe becomes defective and it becomes necessary to remove and replace the same with new material, permit required
Mechanical, replacement of parts	Does not alter approval of equipment or make it unsafe
Painting or similar finish work	None
Papering or similar finish work	None
Pipes, valves, or fixtures—clearing of stoppages or the repairing of leaks	As long as repairs do not involve or require the replacement or rearrangement of valves, pipes, or fixtures
Portable fuel-cell appliances	Not connected to a fixed piping system and not interconnected to a power grid
Public service agencies	Installation, alteration, or repair of generation, transmission, distribution, or metering or other related equipment under the ownership or control of public service agencies by established right (IRC R105.2.3)
Refrigerator system, self-contained	Containing 10 pounds (4.54 kg) or less of refrigerant and actuated by motors of 1 horsepower (746 W) or less
Repairs to structures, ordinary	Does not include cutting away of any wall, partition, or portion thereof, the removal or cutting of any structural beam or load-bearing support, or the removal or change of any required means of egress, or rearrangement of parts of a structure affecting the egress requirements, nor addition to, alteration of, replacement, or relocation of any water supply, sewer, drainage, drainage leader, gas, soil, waste vent, or similar piping, electrical wiring, or mechanical or other work affecting public health or general safety (IRC R105.2.2)
Retaining walls	Not over 4 feet (1,219 mm) measured from the bottom of the footing to the top of the wall, unless supporting a surcharge
Sidewalks and driveways	None
Swimming pools	Prefabricated that are less than 24 inches (610 mm) deep
Swings and other playground equipment	None
Tiling	None
Ventilation appliances, portable	None
Water closets, removal and reinstallation	Provided repairs do not involve or require the replacement or rearrangement of valves, pipes, or fixtures

continues

TABLE 3-1   Work Exempt from Permit—International Residential Code.   *continued*

TYPE OF WORK	SPECIFIC REQUIREMENTS
Water piping, steam, hot or chilled	Within any heating or cooling equipment regulated by code
Water tanks supported directly on grade	If the capacity does not exceed 5,000 gallons (18,925 L) and the ratio of height to diameter or width does not exceed 2 to 1
Window awnings	Supported by an exterior wall which do not project more than 54 inches (1,372 mm) from the exterior wall and do not require additional support

As you can see, if you do minor repair work, such as replacing a lamp, or routine decorating, like painting or papering, you will not need a permit. But if you do structural work, such as removing a wall, you will have to apply for a permit. Sometimes the size of the project determines when you need a permit. For example, if you want to build a tool shed, you don't need a permit under the IRC code if it is smaller than 120 square feet. If it's larger than that, you'll need a permit.

The International Building Code also describes work that does not require a permit. These exceptions are described in Table 3-2.

TABLE 3-2   Work Exempt from Permit—International Building Code.

TYPE OF WORK	SPECIFIC REQUIREMENTS
Accessory structures	Used as tool and storage sheds, playhouses, and similar uses if floor area does not exceed 120 square feet (11.15 m^2)
Cabinets	None
Carpeting	None
Cases, counters, and partitions (movable)	Not over 5 feet 9 inches (1,753 mm) in height
Cooling unit, portable	None
Countertops	None
Electrical equipment for radio and television transmission	But not equipment and wiring for power supply and the installation of towers and antennas
Electrical repairs and maintenance	Minor work, including replacement of lamps or connection of approved portable electrical equipment to approved permanently installed receptacles
Emergency replacement of equipment and emergency repairs	Must submit permit application within the next working business day
Evaporative cooler, portable	None
Fences	Not over 6 feet (1,829 mm) high
Gas, replacement of parts	Does not alter approval of equipment or make it unsafe

TABLE 3-2   Work Exempt from Permit—International Building Code.   *continued*

TYPE OF WORK	SPECIFIC REQUIREMENTS
Heating appliances, portable	None
Leaks in drains and water, soil, waste, or vent pipe; stopping of water, soil, waste, or vent pipe	If any concealed trap, drain pipe, or water, soil, waste, or vent pipe becomes defective and it becomes necessary to remove and replace the same with new material, permit required
Mechanical, replacement of parts	Does not alter approval of equipment or make it unsafe
Oil derricks	None
Painting	None
Papering	None
Pipes, valves, or fixtures (clearing of stoppages or the repairing of leaks therein)	Provided such repairs do not involve or require the replacement or rearrangement of valves, pipes, or fixtures
Public service agencies	Installation, alteration, or repair of generation, transmission, distribution, or metering, or other related equipment under the ownership or control of public service agencies established by right
Refrigerator system, self-contained	Containing 10 pounds (4.54 kg) or less of refrigerant and actuated by motors of 1 horsepower (746 W) or less
Repairs to structures, ordinary	Does not include cutting away of any wall, partition, or portion thereof, the removal or cutting of any structural beam or load-bearing support, or the removal or change of any required means of egress, or rearrangement of parts of a structure affecting the egress requirements, nor addition to, alteration of, replacement, or relocation of any standpipe or water supply, sewer, drainage, drainage leader, gas, soil, waste, vent or similar piping, electrical wiring, or mechanical or other work affecting public health or general safety
Retaining walls	Not over 4 feet (1,219 mm) measured from the bottom of the footing to the top of the wall, unless supporting a surcharge or impounding Class I, II, or III-A liquids
Sets and scenery	Temporary motion picture, television, or theater stage sets and scenery
Shade cloth structures	For nursery or agricultural purposes and not including service systems
Sidewalks and driveways	No more than 30 inches (762 mm) above grade and not over any basement or story below and which are not part of an accessible route
Swimming pools	Prefabricated and accessory to a Group R-3 occupancy, which are less than 24 inches (610 mm) deep, do not exceed 5,000 gallons (18,925 L), and are installed entirely above ground
Swings and other playground equipment	Accessory to detached one- and two-family dwellings
Testing systems, temporary	Required for testing or servicing of electrical equipment
Ventilation equipment, portable	None
Tiling	None
Water closets, removal and reinstallation	Provided repairs do not involve or require the replacement or rearrangement of valves, pipes, or fixtures

continues

TABLE 3-2 Work Exempt from Permit—International Building Code. *continued*

TYPE OF WORK	SPECIFIC REQUIREMENTS
Water piping, steam, hot or chilled	Within any heating or cooling equipment regulated by code
Water tanks supported directly on grade	If the capacity does not exceed 5,000 gallons (18,925 L) and the ratio of height to diameter or width does not exceed 2 to 1
Window awnings	Supported by an exterior wall which do not project more than 54 inches (1,372 mm) from the exterior wall and do not require additional support of Group R-3 as applicable in Section 101.2 and Group U occupancies

Figure 3–3 Carpet installation is exempt from a permit.

Figure 3–4 Painting your house is exempt from a permit.

Figure 3–5 A carpenter making cabinets for home installation is exempt from a permit.

These tables are for the model ICC codes. Model codes can be amended by a state or town, so the only way to make sure you do not need a permit for an activity is to check the building code of the local jurisdiction.

Checking With The Building Department

While you may find out the answer to the permit question by calling the local building official, you have to make sure you are speaking to a person who has the authority to answer the question properly and not someone who is just answering the telephone. For example:

Caller: I'm doing some work around the house and I want to know if I need a permit to replace some tile in my bathroom.

Building Inspector: "No, that's just a repair and you don't need a permit."

The reality: The caller was putting in a shower stall with a new wall he intended to tile, all of which required a permit.

Case point: You'll only get an answer based on the information you give. If you give incomplete information,

Figure 3–6 A contractor installing tile backsplash in a kitchen is exempt from a permit.

Figure 3–7 The construction of a shower stall with a new wall must have a permit.

you won't get the information you need to prevent serious implications down the road. The person most affected will be you.

People create problems for themselves by calling the building department and giving incomplete information. Give complete information to someone with the authority to give you an answer when you ask whether you need a permit.

A common excuse in court from people who are charged with working without a permit is "the city told me I didn't need a permit." When they are asked by the prosecutor whom they spoke to, the usual answer is "the lady at the reception desk told me. . ." This type of interaction is almost impossible to trace. Even if the

Figure 3–8 When asking for guidance from a building authority on whether or not a permit is needed, it is a good idea to get the person's name or, preferably, an answer in writing.

JUST ✔ CHECKING

Work Exempt From Permit Requirements

- ☐ Does the building code cover the type of activity to be performed?
- ☐ Is the activity to be performed exempt from the requirements of the code?
- ☐ What section number of the building code is being relied upon to base that exemption?
- ☐ Has the building department been called to make sure the work is exempt?
- ☐ Did the person consulted in the building department have the authority to render an opinion?
- ☐ What information was provided to the building department about the work to be performed at the time the person was asked to render an opinion regarding the exemption?

---

prosecutor is able to find out who the "lady" is, her recollection of the conversation is usually quite different. The solution to this dilemma is: When in doubt about whether you need a permit, get it in writing or at least get the name of the person having authority who made the decision.

Exceptions To The Fire Code

Other codes contain different exceptions to the requirements for a permit. The International Fire Code does not cover the buildings and activities listed in Table 3-3. This doesn't mean that the activity is not covered by a code, just that the fire code doesn't cover it.

TABLE 3-3 Exceptions to the Requirements of the International Fire Code.

EXCEPTION	IFC SECTION NUMBER
Change in use or occupancy of existing structure with the approval of the fire code official	102.3
Historical buildings—construction, alteration, repair, enlargement, restoration, relocation or moving	102.5

AVOIDING TROUBLE

- Always check the code of the local jurisdiction to see if the work you are doing is exempt from the requirements of the code.
- Don't rely on your own interpretation if there is any question about the work being exempt.
- Contact someone in a position of authority at the local jurisdiction to give you an answer as to whether the work is exempt.
- Get the answer from the local jurisdiction in writing.
- Don't forget to check fire code requirements too.

CONCLUSION

It is important to determine whether the work you are doing needs a permit. Each building code describes in detail what work is exempt from the requirements of a permit. Do not rely on your own interpretation if there is any room for guessing. Call the building department and speak with someone who has the authority to give you a definitive answer.

CHAPTER 4

PERMIT APPLICATION

INTRODUCTION

If you do need a permit, you will have to fill out an application. You must give all the required information and it must be accurate. One of the most time-consuming tasks you will undertake is putting together the information and documents for the application. This chapter will take you through a sample application and explain what types of information you need.

CONTENTS OF A BUILDING PERMIT APPLICATION

The following sample form is an example of what a typical building application may look like. You should be prepared to have the information required on the building application for the jurisdiction in which you want to build. It does look very complicated, but think of it in small sections. The building department needs to know who is pulling the permit and where the property is. It needs to know what you propose to build so it can determine if you need a permit, and if so, what code applies. It needs to know what the zoning district is so that it can determine if this type of construction is allowed in it. The application is illustrated with comments to help you understand why you are being asked for certain information. Fill out the application completely before turning it in to the building department. This will save you time because it won't be rejected because of missing information.

KEY CONCEPTS

building permit application—a written request that an official document be issued by the authority having jurisdiction giving permission to a particular person or entity for the performance of a specific activity.

Figure 4–1   One of the most time-consuming tasks you will undertake is putting together the information and documents for your permit application, but it is important to fill it out correctly.

Figure 4–2   Sample building permit.

Every building department uses its own application, and many will not be as thorough as this one. You should obtain the form ahead of time so you know what type of information you will need to properly fill it out.

VILLAGE OF WOODRIDGE
Department of Building Safety
Three Plaza Drive
Woodridge, IL 60517
APPLICATION FOR PERMIT

For office use only: ....

APPLICATION NUMBER: _____ PERMIT NUMBER (if different): _____

PERMIT FEE: _____ DATE RECEIVED: _____ TIME RECEIVED: _____

REVISIONS RECEIVED: _____

Property Address: ....

Property Index Number: _____

Legal Description: ....

Lot Dimensions: Frontage: _____ Depth: _____ Square Feet: _____

Zoning District: ....

Existing Uses: ⊡ Single Family ⊡ Multifamily ⊡ Commercial/Industrial ☐ Public

☐ Retail/Office ☐ Agricultural ☐ Institutional ☐ Other: _____

Proposed Use: _____

Check *all* applicable items:

BUILDING	BUILDING (MINOR)	ACCESSORY
☐ New construction	☐ Roof	☐ Accessory structure
☐ Addition	☐ Fence/wall	☐ **Paving**
☐ Alteration/Remodel	☐ Chimney	☐ Parking lot
☐ Conversion	☐ **Sign**	☐ Parking lot improvement
☐ Foundation only	☐ Type (wall/pole)	☐ Front yard paving/driveway
☐ Unreinforced masonry	☐ Fixtures (Qty)	**FIRE PERMITS**
☐ After the fact permit	☐ Incandescents (Qty)	☐ Fire suppression system
☐ Movement of structure	☐ Ballasts (Qty)	☐ Fire alarm
☐ **Demolition**	☐ Transformers (Qty)	☐ Sprinklers
☐ Full/partial	☐ **Pool**	☐ Underground sprinklers
☐ **Grading**	☐ Public/private	☐ Monitors
☐ Hillside/Non-hillside	☐ Elect fixtures (qty)	**MISCELLANEOUS**
☐ Solar	☐ Motor less than 1 hp	☐ Spray booth
☐ Tenant improvement	☐ Motor less than 5 hp	☐ Fuel storage tank
☐ Conditional	☐ Pool heater	☐ Range hood
☐ Repair	☐ Backwash disposal	☐ Satellite dish antenna
☐ Other _____	☐ **Elevator or chair lift**	☐ Lawn sprinkler system

This is for the use of the jurisdiction, and you don't have to fill out anything that says "for office use only."

The following lines describe the property so there is no mistake as to where the job is to be performed.

This will be on the deed that transferred the property to the owner and the title insurance report issued at the time of the sale.

This helps the building department determine whether the lot is big enough for the job.

The zoning district identifies the type of structures that are allowed on the lot (e.g., you can't build a gas station in a residential zoning district).

This tells the building department what the property is currently being used for.

The proposed use must be described so that the building department can determine if the use is allowable. If there is a change of use, other code regulations may have to be followed.

With every box that applies checked, the building department will determine the kind of permit, the scope of the permit, and how many permits you need.

Your project may require fire safeguards. This will add considerable cost to the project, but you will not be able to use the structure if you do not comply with the fire code.

Figure 4–3 Sample permit application form, International Building Code.

Description of Work: ....

_____

Flood plain: ☐ Yes   ☐ No·······································

Value of construction: _____ Cubic feet: _____

For additions: Existing square feet: ············· Proposed addition: _____

Owner: _____

Address: _____ City, State: _____

Zip: _____

E-mail: _____ Office #: _____ Cell: _____

FAX: _____

Check appropriate boxes and fill in the information requested:

☐ Individual—D.O.B. _____ D.L. No. _____

☐ Corporation—Corporate No. _____   ☐ Limited Liability Company—LLC No. _____

Registered Agent: _____

Address: _____

City, State: _____ Zip: _____

☐ Land Trust, Trustee: _____

Person with Power of Direction: _____

Address: _____

City, State: _____ Zip: _____

Beneficiaries: _____

Address: _____

City, State: _____ Zip: _____

☐ Other: _____

Address: _____

City, State: _____ Zip: _____

Design Professional: ··································

Address: _____ City, State: _____ Zip: _____

Office #: _____ Cell: _____ FAX: _____

State License #: _____ E-mail: _____

General Contractor: ································

Address: _____ City, State: _____ Zip: _____

Office #: _____ Cell: _____ FAX: _____

E-mail: _____

Check appropriate boxes and fill in the information requested:

☐ Individual—D.O.B. _____ D.L. No. _____

☐ Corporation—Corporate No. _____   ☐ Limited Liability Company—LLC No. _____

It is very important to clearly describe the type of work you are performing, otherwise you could get in trouble for working beyond the scope of the permit.

Different types of code provisions apply if you are working in a flood plain area. You may also need a very specific permit called a stormwater management permit.

This information is required for the local taxing authority so it can place a value on the building for property tax purposes.

The building department needs this information to determine if the addition is too large for the property.

The building department wants to know who owns the property in case something goes wrong during construction so it can contact the owner and try to fix the problem. Also, since the building inspectors are going on the property, they want to make sure that the owner knows they will be coming on the property to perform inspections. Specific information on the legal entity that owns the property is important information for the building department to have so that if something goes wrong, the building official can deal with the proper party.

If the legal entity applying for the permit is a corporation or LLC, have all of the information about it with you when you apply. The information will be in your Articles of Incorporation or annual report of the corporation or LLC.

This includes living trusts in which the trustee is also the beneficiary and person with the power of direction.

If there is a design professional, it is helpful to have the contact information on file.

If there is a general contractor, it is vital to have the contact information on file.

Figure 4–3   *continued*

Registered Agent: _____

Address: _____

City, State: _____ Zip: _____

☐ Other: _____

Address: _____

City, State: _____ Zip: _____

License #s:

Contractor _____ Roofing Contractor _____

Plumbing _____ Electrical _____

Other _____

Applicant (if different from owner) ·····················

Name: _____

Address: _____ City, State: _____ Zip: ____

Office #: _____ Cell: _____ FAX: _____

E-mail: _____

Check appropriate boxes and fill in the information requested:

☐ Individual—D.O.B. _____ D.L. No. _____

☐ Corporation—Corporate No. _____   ☐ Limited Liability Company—LLC No. _____

Registered Agent: _____

Address: _____

City, State: _____ Zip: _____

☐ Other: _____

Address: _____

City, State: _____ Zip: _____

> The permit holder is responsible for the work being done on the property, so the building department wants complete information on the applicant in case something goes wrong.

SUBCONTRACTOR INFORMATION ··············

Subcontractor: _____

Address: _____ City, State: _____ Zip: ____

Office #: _____ Cell: _____ FAX: _____

E-mail: _____

Check appropriate boxes and fill in the information requested:

☐ Individual—D.O.B. _____ D.L. No. _____

☐ Corporation—Corporate No. _____   ☐ Limited Liability Company—LLC No. __

Registered Agent: _____

Address: _____

City, State: _____ Zip: _____

License #: _____

UNDER PENALTY OF INTENTIONAL MISREPRESENTATION AND/OR PERJURY, I declare that I have examined and/or made this application and it is true and correct to the best of my knowledge and belief. I agree to construct said improvement in compliance with all

> Some towns closely supervise the job site and fine violators. Knowing who the subcontractors are before construction begins assists the building official in issuing violation notices to the correct party.

> The building department wants to make sure that the information on the application is true and correct. Therefore, there is a penalty for people who lie on their application. This is very serious and can lead to a criminal prosecution if the applicant deceives the building department.

Figure 4–3   *continued*

provisions of the applicable ordinances. I further certify that all easements, deed restrictions, or other encumbrances restricting the use of the property are shown on the site plans submitted with this application. I have been given authorization from the property owner to obtain this permit. I realize that the information that I have affirmed hereon forms a basis for the issuance of the permit herein applied for and approval of plans in connection therewith shall not be construed to permit any construction upon said premises or use thereof in violation of any applicable ordinance or to excuse the owner or his or her successors in title from complying therewith.

_____   _____

Applicant's Signature and Date   Title

This is where you sign the application.

I understand that by applying for this permit, I am consenting to the inspection of this property and to the entry onto the property by inspectors of the authority having jurisdiction for the purpose of performing the necessary inspections during normal business hours for the duration of the permit.

_____

Owner's Signature

If you are the owner or agent for the owner, this is where you would fill in the correct title.

The building department wants to make sure the owner consents to the inspections that are necessary.

Office Use Only
APPROVALS:
Type   **Date**   Initials
Building
Electrical
Mechanical
Plumbing
Concrete
Engineering
Water
Sewer
Fire
Health
Landscaping
Energy
Medical Gas
Zoning
Historical
Special Inspections
Site Plan
Flood Plain Yes__ No__

The owner should sign here, acknowledging that he or she consents to the inspection.

This is a space for the building department to approve the work that has been performed.

These are the different types of inspections that may be required by the permit.

Figure 4–3 *continued*

To help you make sure you have enough information to submit the permit application, use the following checklist.

JUST ✔ CHECKING

Permit Applications

- ☐ Does the type of work described require a permit?
- ☐ Have I verified with the building department that I need a permit?
- ☐ Is more than one type of permit required?
- ☐ If so, what type of additional permit is needed?
- ☐ Do I need the approval of the fire district or department for the work being done?
- ☐ Does any of the work described fall under an exception to the code requirements for a permit?
- ☐ Have I checked with the building department to make sure I don't need a permit and that the work meets the requirements for an exception to the code?
- ☐ Have I identified and described the work to be covered by the permit in the application?
- ☐ Have I identified the land on which the proposed work is to be done so it can be located by:
  - ☐ Legal description?
  - ☐ Street address?
  - ☐ Similar description?
  - ☐ Property index number?
- ☐ Have I identified the use and occupancy for the proposed work in the application?
- ☐ Do I have the necessary construction documents and other information required by the code to accompany the application?
- ☐ Is the valuation of the proposed work on the application?
- ☐ Is the application signed as required?
- ☐ Is the application complete?
- ☐ Is the building or structure located in an area prone to flooding?
- ☐ If so, are there special requirements that I need to address?

continues

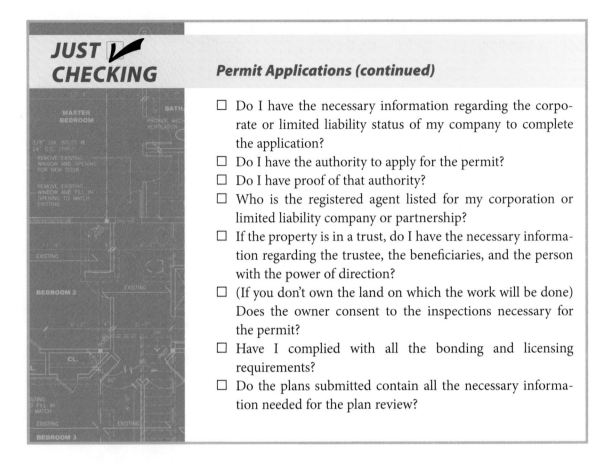

JUST ✔
CHECKING

Permit Applications (continued)

- ☐ Do I have the necessary information regarding the corporate or limited liability status of my company to complete the application?
- ☐ Do I have the authority to apply for the permit?
- ☐ Do I have proof of that authority?
- ☐ Who is the registered agent listed for my corporation or limited liability company or partnership?
- ☐ If the property is in a trust, do I have the necessary information regarding the trustee, the beneficiaries, and the person with the power of direction?
- ☐ (If you don't own the land on which the work will be done) Does the owner consent to the inspections necessary for the permit?
- ☐ Have I complied with all the bonding and licensing requirements?
- ☐ Do the plans submitted contain all the necessary information needed for the plan review?

Who Should Apply For The Permit?

The other decision you must make is whether the owner of the property or the contractor will apply for the permit. In many cases, the contractor acts as the agent for the owner and obtains the permit. It does not really matter who applies for the permit as long as somebody does, unless the local jurisdiction has special rules. Always make sure you have a permit before the work begins. The building department will give you a permit placard that you must usually put up in a visible place on the job site. You can actually receive a notice of violation if the permit placard is not property posted. If you have this permit placard, you know that the permit was approved.

This is what a typical permit looks like:

VILLAGE OF HINSDALE
BUILDING AND USE PERMIT

> This tells you where you need to display the permit.

NOTICE: This must be displayed in a conspicuous place, facing the street, on or near the building for which it is issued before starting the work, and remain posted until the work is completed.

> This shows that the permit has been approved by the building department.

APPROVED BY: _____ DATE: _____ ISSUED BY: _____

Application Number: _____ Date: _____

> This should correspond with your permit application number.

Property Address: _____ Property Index Number: ___

> This identifies the location of the property.

Temporary Structure Permit: Yes ☐   No ☐

> This identifies whether the structure is temporary.

Expiration Date: _____

> This tells you when the permit expires. You may need to apply for an extension if the permit expiration date is near.

Item	Inspector	Date	Result
Footing			
Foundation			
Backfill			
Spot Survey			
Sewer and Water			
Electrical Service			
Electrical Rough			
Plumbing Rough-In			
Plumbing			
Underground			
Framing			
Insulation			
Basement, Garage			
Slab Floor			
Sidewalk			
Topographical			
Survey			
Final Building and			
Plumbing			
Final Electrical			
Final Engineering			
Occupancy			

> This is the section used by the inspector to identify what inspections have been completed along with the result.

> This is general information about obtaining inspections.

TO CALL FOR INSPECTIONS, CONTACT THE DEPARTMENT OF BUILDING SAFETY AT *630-555-7030*. THIS PERMIT AUTHORIZES ONLY WORK IN FULL COMPLIANCE WITH VILLAGE CODES AND REGULATIONS. FAILURE TO OBTAIN INSPECTIONS AS REQUIRED CONSTITUTES A VIOLATION OF THE RESIDENTIAL CODE.

(Reverse side of permit)

Figure 4–4   Sample building permit.

Property Zoning: _____ Application Type Description: _____

Subdivision Name (if applicable) _____ Application Valuation _____

This is general information taken from the permit application and is filled out by the building department.

Owner:

Name: _____

Address _____

City, State, Zip: _____

Phone No. _____

Contractor:

Name: _____

Address: _____

City, State, Zip: _____

Phone No. _____

Permit Type: BUILDING PERMIT

This section identifies the type of permit issued along with specific information about it.

Additional Description:

Permit Fee:

Issue Date:

Expiration Date:

Permit Type: ELECTRICAL PERMIT

Additional Description:

Permit Fee:

Issue Date:

Expiration Date:

Permit Type: PLUMBING PERMIT

Additional Description:

Permit Fee:

This section is filled out by the building department and contains information regarding the cost of the permit, the date it is issued, and the date it expires.

Issue Date:

Expiration Date:

Other Fees: Review—SFR _____

If there are other types of fees, the amount will be listed here by the building department.

Sewer Connection _____

Unmetered Water _____

Water Connection _____

Water Meters _____

Fee Summary	Charged	Paid	Credited	Due
Permit Fee Total				
Other Fee Total				
Grand Total				

Construction may require fire department inspections and tests.

Item	Inspector	Date	Result
Fire Alarm Test			
Sprinkler System Test			
Final Inspections			

If fire code inspections are required, this section will be used by the fire inspector.

Figure 4–4 *continued*

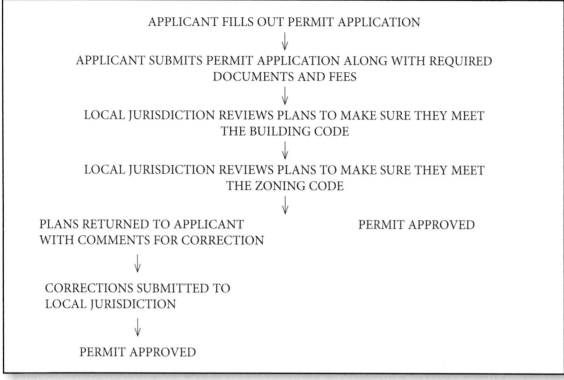

APPLICANT FILLS OUT PERMIT APPLICATION
↓
APPLICANT SUBMITS PERMIT APPLICATION ALONG WITH REQUIRED
DOCUMENTS AND FEES
↓
LOCAL JURISDICTION REVIEWS PLANS TO MAKE SURE THEY MEET
THE BUILDING CODE
↓
LOCAL JURISDICTION REVIEWS PLANS TO MAKE SURE THEY MEET
THE ZONING CODE
↓

PLANS RETURNED TO APPLICANT   PERMIT APPROVED
WITH COMMENTS FOR CORRECTION
↓
CORRECTIONS SUBMITTED TO
LOCAL JURISDICTION
↓
PERMIT APPROVED

Figure 4–5   Permit application procedure.

Requirements For Corporations, Limited Liablity Companies, Or Partnerships

If your business is a corporation or a limited liability company or partnership, you will need your company information for the application. It is common for applicants to list insufficient information for a company. An applicant might fill in the name "M&E Construction" for the contractor on the form. This name does not contain anything that would indicate that it is a legal corporation or limited liability company or partnership, or a properly registered assumed name. It is impossible to tell what the legal entity is. You should use the exact name of the company registered with the state on your incorporation papers. When you applied to become a corporation, limited liability company, or partnership, your name had to contain something identifying it as an incorporated entity, such as "Corporation," "Incorporated," "Limited," or "Company," or abbreviations such as "Corp.," "Inc.," "Ltd.," "Co.," "L.L.C.," or "L.L.P." You must use the full name of your company, for example, M&E Construction Company. If you use a business name but own the company as a sole proprietor, you must use your name followed by your business name, for example. Larry Jones d/b/a Jones Construction. If your company is a corporation or a limited liability company, make sure you make that clear on the permit application.

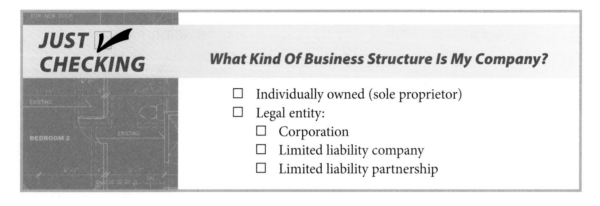

JUST ✔ CHECKING

What Kind Of Business Structure Is My Company?

- ☐ Individually owned (sole proprietor)
- ☐ Legal entity:
  - ☐ Corporation
  - ☐ Limited liability company
  - ☐ Limited liability partnership

Multiple Permits

Very often, more than one type of permit is required. The building department will let you know if this is required based on the plans you submit. This is why the application must show all the work you want to do. The building department cannot give you accurate information regarding the permits needed if you withhold information.

If there will be any change to the grade of the property, you will usually have to obtain a grading permit. This will often require a topographical survey that will show elevations and how water drains on the property. If you are in a wetland or flood plain, there

Figure 4–6 Residential buildings along wetlands, such as this, have special requirements because of the building's proximity to water and wildlife.

may be other special requirements, such as an engineering study along with a stormwater management permit. This is often overlooked by builders and homeowners and can get you in very big trouble with the agency that oversees stormwater management issues. Always check with the building department to make sure you have all of the permits required. For more information on these kinds of issues, see Chapter 2.

JUST ✔ CHECKING

Necessary Information On An Application

What's needed for the application?*
- ☐ Identity and description of work to be covered by the permit
- ☐ Description, street address, or similar description that will readily identify and definitely locate the proposed work or building
- ☐ Use and occupancy for which the proposed work is intended
- ☐ Construction documents and other information required by the code
- ☐ Valuation of the proposed work
- ☐ Signature of the applicant or applicant's authorized agent
- ☐ Other data and information required by the building official

* From the International Residential Code

Permit Technicians

Many local jurisdictions hire permit technicians whose job it is to assist applicants with the permit process. They are the people who really know the ins and outs of processing the paperwork. They can be extremely helpful to you. Though they cannot fill out the application for you, they will try to guide you throughout the permit application process. Having a mutually respectful relationship with the local permit technician can make the application process a pleasure instead of a chore.

Figure 4–7 Many local jurisdictions have permit technicians whose job it is to assist applicants with the permit process. Developing a mutually respectful relationship with these persons can ensure a smoother application process.

AVOIDING TROUBLE

- Obtain a blank permit application form so you know what kind of information you need.
- Gather all of the necessary information before filling out the form.
- If you are a corporation or other legal entity, have all of the legal documents necessary to complete the application.
- Do not submit an incomplete application.
- If you are confused, ask a permit technician for assistance.

CONCLUSION

Having the necessary information before applying for a permit will help you sail through the application process. Gather the necessary information ahead of time so you don't have your application rejected because it is incomplete. Also, make sure you apply for all the permits you need. If you need assistance with this, speak with the staff at the building department for guidance. Get to know the permit technician handling your application. It helps to have someone who will try to help you should you run into any difficulties with the process.

CHAPTER 5

CONSTRUCTION DOCUMENTS

INTRODUCTION

There are specific requirements regarding what type of construction documents must be submitted with the permit application depending on the type of permit you are seeking. Many cities have a handout or guide available regarding what those requirements are. Most municipalities post these requirements on the town's website. If you need to apply for a permit, begin by checking the website for that information before doing anything else. If the information is not on the website, obtain a copy of the city's printed guidelines. If you don't have the right documents or a complete application, you may delay your project and not get a permit.

CONSTRUCTION DOCUMENT REQUIREMENTS

Most cities require at least two sets of plans or drawings with specifications and calculations that meet the architectural, mechanical, structural, and electrical requirements of the building code. Certain states require that the plans be prepared by a registered design professional. If there are unusual circumstances, additional plans prepared by a registered design professional may be required.

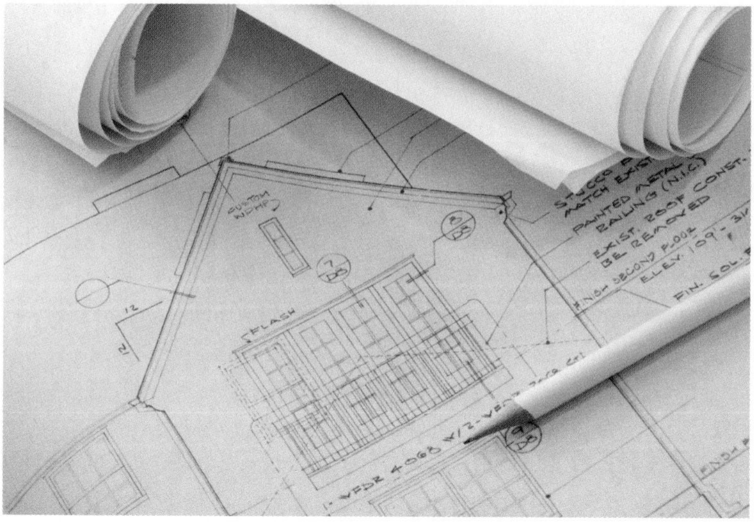

Figure 5–1 Some states require that plans be prepared by a registered design professional. Unusual circumstances can sometimes necessitate additional professional plans.

KEY CONCEPTS

survey—the process by which a parcel of land is measured and its contents ascertained.

Black's Law Dictionary

KEY CONCEPTS

site plan—blueprint and detailed plan for the development and intended use of a particular piece of land.

BusinessDictionary.com

The most common mistakes made by permit applicants are:

- Submitting incomplete applications
- Submitting information that is too general about the project
- Listing incomplete contractor information (e.g., no license, insurance, or bond information)
- Not submitting a plat of survey or one that is obsolete
- Not submitting copies of contracts or proposals for work

Hire a registered design professional if required for your project. He can help you avoid delays and rejection of your construction plans.

Construction documents are supposed to be drawn on suitable material. Many building departments accept electronic media documents. The construction documents must be sufficiently clear and show the location, nature, and extent of the work proposed. There must be enough detail to show that the work will meet the requirements of the building code and any other applicable laws.

Other Required Documents

Copies of a plat or survey of the lot showing the location of the construction are often required. A survey shows the boundary lines on the property along with easements and setbacks. For a further description as to why these matter, see Chapter 2.

A site plan is a common requirement because it shows the size and location of the new construction and the existing structures on the site and the distances from the lot line. If a structure on the property is going to be demolished, the site plan will need to show what is to be torn down, and the location and size of the existing structures and buildings that are going to remain on the property.

Certificates of insurance, bonds, and/or licenses must be submitted with the application. If you are a contractor and your project is going to involve the use of public streets or other public property, you will probably have to file a certificate of insurance showing that you have the necessary amount of public liability and property damage insurance required by the local jurisdiction. The town may require a surety bond payable to it in case there is damage to the public property. During a construction project, very few things go perfectly. Sewer lines are accidentally broken, streets may need repair after the building's sewer line is connected to the public one, and parkway trees may die as a result of construction damage.

In the Field

"We are very protective of trees on the town parkway. We're an old town so the trees are large. Replacing them can cost the contractor a lot if they are damaged and the town isn't going to pay for the cost to plant a new tree. We often have contractors who don't get their full bond back because they damaged our trees." —*Robb, Building Official* ■

Required Licenses

If the state or local law requires that an electrician, plumber, roofer, or contractor be licensed, the municipality will require proof of that license by way of a certification from the state and/or a copy of the photo identification of the person holding the license.

Many building departments send out "welcome letters" to new permit holders. That way, if someone has illegally used a contractor's state license number without his permission, the real contractor will call the building department if the applicant didn't apply for the permit. Building departments often require a photo identification card from anyone applying for a permit so that the building official knows who the person is if someone lies on the application. Unfortunately, building departments have instituted these policies because many unqualified people will try to get a permit by providing false information.

Figure 5–2 Damage to parkway trees during construction is considered damage to public property.

Property In Flood Hazard Areas

If the property is located in a flood hazard area, there are special requirements for documents that must accompany the permit application. For example, the International Residential Code requires the following:

- Delineation of flood hazard areas, floodway boundaries, and flood zones and the design flood elevation as appropriate
- Elevation of the proposed lowest floor, including basement, in areas of shallow flooding (AO zones), the height of the proposed lowest floor, including basement, above the highest adjacent grade
- Elevation of the bottom of the lowest horizontal structural member in coastal high-hazard areas (V Zone)
- Design flood elevation and floodway data available from other sources if design flood elevations are not included on the community's Flood Insurance Rate Map (FIRM)

Figure 5–3 If a property is located in an area prone to flooding, special documents may need to be in your permit application.

AFTER THE PLANS ARE SUBMITTED

After the plans are submitted, a plan review examiner will review them. If they are approved, the building department will do so in writing or by stamping the construction documents. Usually one set will be kept by the building department and the other returned to you. You should keep a copy of the approved plans on the job site

Figure 5–4 A building official may inspect plans on the job site.

along with manufacturer's installation instructions. The building official may inspect them on the job site.

Once your plans are approved and the permit is issued, no changes should be required by the building code during the period of construction as long as it is being pursued in good faith. If you let your building permit lapse and the code changes during that time, you may have to get new plans to comply with the updated building code.

Use the following checklist to determine if you have all the information you need before you submit your permit application.

Construction Documents

JUST ✔ CHECKING

☐ Have the proper number of sets of the construction documents been submitted?

☐ Have the construction documents been prepared by a registered design professional (if required by state statutes or ordinances of the local jurisdiction)?

☐ Are there any additional construction documents that should be prepared by the design professional?

continues

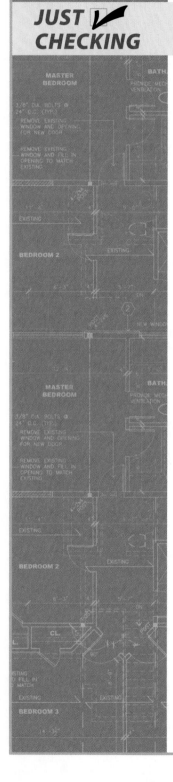

JUST ✔ CHECKING

Construction Documents (continued)

☐ Should you ask the building official to waive the submission of construction documents or other data, not required to be prepared by a registered design professional, because the nature of the work applied for is such that reviewing the documents is not necessary to obtain compliance with the code?

☐ Will the building official allow submission of electronic media documents?

☐ Are the construction documents sufficiently clear to show:

    ☐ The location of the work?

    ☐ The nature of the work proposed?

    ☐ The extent of the work proposed?

    ☐ That the work will conform to the provisions of the relevant laws, ordinances, rules, and regulations?

☐ Are manufacturer's installation instructions required by the code available on the job site?

☐ Do the construction documents for buildings and structures located in whole or in part in flood hazard areas include the following:

    ☐ Delineation of flood hazard areas, floodway boundaries, and flood zones and the design flood elevation as appropriate?

    ☐ Elevation of the proposed lowest floor, including basement, in areas of shallow flooding (AO zones), the height of the proposed lowest floor, including basement, above the highest adjacent grade?

    ☐ Elevation of the bottom of the lowest horizontal structural member in coastal high-hazard areas (V Zone)?

    ☐ Design flood elevation and floodway data available from other sources if design flood elevations are not included on the community's Flood Insurance Rate Map (FIRM)?

☐ Do the construction documents include a site plan showing the size and location of the new construction

Construction Documents (continued)

JUST ✔ CHECKING

and existing structures and constructions that are to remain on the site or plot?

☐ If the application is for demolition, does the site plan show the construction to be demolished and the location and size of existing structures and constructions that are to remain on the site or plot?

☐ If the construction documents have been approved, does the approval appear via writing or a stamp on the documents?

☐ Has a set of the approved plans been returned to you?

☐ Has the building official retained a copy of the approved plans?

☐ Is a set of the approved plans kept at the site of the work and available to the building official and his or her authorized representatives?

☐ Will the building official grant a phased approval for the work; that is, will he or she issue a permit for the construction of the foundations or any other part of the building or structure before the construction documents for the whole building or structure have been submitted?

☐ If so, have adequate information and detailed statements been filed complying with pertinent requirements of the code?

☐ Have amended construction documents for any changes made during construction been submitted for approval?

☐ Has the work been installed in accordance with the approved construction documents?

Phased Approval

In some jurisdictions, you may be able to submit plans for a permit that will allow you to construct the foundation or another part of the building or structure before the construction documents for the whole building or structure have been submitted if adequate statements are filed complying with pertinent requirements of the building code. The building official may be authorized to do this but isn't required to. Also, building officials come and go, and the next person may do things differently. However, most codes provide that you proceed at your own risk with no guarantee that a permit for the entire structure will ever be granted.

Amended Construction Documents

You may not deviate from the approved plans without the permission of the building official during the construction process. While it is not unusual to change plans during construction, you should first submit amended construction documents for approval. Make sure you get written approval for these changes. Verbal approval from an inspector will be worth very little if he or she denies agreeing to the change and the city withholds a certificate of occupancy for the structure. Have the inspector initial the changes on the approved plans. A form changing the construction documents looks like this:

PERMIT NUMBER: .... _____ ····· The number on your permit.

DATE REQUEST FOR CHANGE IN CONSTRUCTION DOCUMENTS RECEIVED:

_____

Property Address: _____   This should be a date before you do the work described in the request for change.

Property Index Number: _____

Change(s) requested: ······· Address of the job site.

_____   Be specific so there is no issue over what is being changed

_____

_____   This is where you sign the request for change.

_____

Applicant's Signature and Date   Title   If you are the owner or agent for the owner, fill in the proper title.

Approved: _____ Date: _____   This is where the building official signs if the change is approved and the date of the approval.
　　　　Building Official's Signature

Approved as noted: _____ Date: _____
　　　　　Building Official's Signature   This is where the building official signs if the change is approved but with conditions.

Figure 5–5 Any changes to approved plans must be approved by the building official during the construction process. It's not unusual for plans to change, but you must submit amended construction documents and get written approval from your building official.

Denied: .... Date: ....
            Building Official's Signature

This is where the building official signs if the change is denied. If so, you may not change the plans.

Amended construction documents submitted: Yes ☐ No ☐ ··············

Check the "yes" box if you submit additional construction documents as part of the change request and the "no" box if there are no additional documents.

Notes: _____

_____

These are notes by the building official that describe any conditions or changes he makes regarding your request for a change in the construction documents.

Figure 5–5 *continued*

AVOIDING TROUBLE

- Make sure your construction documents are in order before you submit them.
- If you are in a flood hazard zone, make sure you submit all necessary plans.
- If you need to change your plans, get approval in writing from the building inspector or other necessary person.
- If you are using a phased document process, don't assume you'll obtain all the necessary approval in the future.

CONCLUSION

By submitting the necessary construction documents in the proper form, you can be confident that the permit process will go smoothly. If you need to change your plans in the middle of construction and you get permission in writing from the building official, you will ensure that completion of the job is not held up.

CHAPTER 6

INSPECTIONS

INTRODUCTION

All building codes require the permit holder to obtain and pass inspections. Most codes subject all construction or work for which a permit is required to inspection by the building official. It is common for the city or town to give out a list of the required inspections when you obtain the permit.

IMPORTANCE OF REQUIRED INSPECTIONS

Inspections play a very important role in making sure that the work performed meets the standards in the building code. The inspector wants to make sure that the work done has been done correctly and safely. It is an opportunity to make sure no changes have been made to the plans and to catch mistakes made in the plan-review process. If a mistake has been made, you must still follow the code in the interests of safety.

Too often people doing the work do not follow the building code and the inspector catches a violation that could lead to a tragedy in the future. People doing construction

Figure 6–1   An inspector wants to make sure that new construction has been done correctly and safely.

vary greatly in their skill and abilities. Inspections verify that the building or work is being done according to the plans submitted with the proper materials.

In the Field

"We once had someone attach sprinkler heads to the ceiling because they were required by the fire code. But, when the inspector did the inspection, he discovered there weren't any pipes connected to them in the ceiling. When he looked into the ceiling, he could not believe it!" —*Scott, Building Official* ∎

Responsibilities Of The Permit Holder

It is the permit holder or permit holder's agent's responsibility to notify the building official when the work is ready for inspection. You must call them; they will not call you. The person requesting the inspection must provide access to and the means for an inspection of the work. This means that you must not drywall the area where the inspector needs to check the plumbing, electrical, or mechanical work. If you do, you will have to tear out a portion of the wall to provide access at your own expense. If you complain about having to do this, the inspector will not be sympathetic because it is not the inspector's fault. Call for inspections as soon as the work is ready so that your construction is not delayed while waiting for the inspector.

Figure 6–2 The person requesting the inspection must provide access to and the means for an inspection of the work. If an inspector cannot, for example, see all electrical work while evaluating an outlet, he may miss something that could be dangerous to inhabitants.

What Happens During The Inspection?

During the inspection, the inspector will look at the work and make sure it is up to code. If the work does not pass the inspection, the inspector will let you know and may give you a written form similar to the one below that tells you what is wrong with the work.

INSPECTION RECORD MEMORANDUM

Permit number: _____ Date of inspection: _____

Address: _____ Time of inspection: _____ a.m./p.m.

Owner/Occupant: _____ Date of inspection notification: _____

Contractor: _____ Call-back number: _____

Type of inspection: _____ Person who called in notice: _____

Inspector assigned: _____

ITEMS TO BE CORRECTED:

_____ _____

_____ _____

_____ _____

_____ _____

_____ _____

_____ _____

_____ _____

_____ _____

_____ _____

☐ **APPROVED**

☐ **NOT APPROVED***

Date: _____

Signature of Building Official/Inspector

*As soon as these corrections are made, contact the Department of Building Safety for a follow-up inspection: 630-555-4750.

This area contains the necessary reference information regarding the permit number, the date and time the inspector performed the inspection, the address of the job site.

This is for the name of the responsible party.

This is the date the contractor or homeowner called the inspector to set up the inspection.

This is for the name of the contractor if there is one.

This should be the telephone number of the person the inspector should contact about any issues.

This is for nature of the inspection, e.g., a final inspection.

This space is for the name of the person who called to set up the inspection.

This is for the name of the inspector who will do the inspection.

The inspector lists all of the items that must be corrected in order to pass the inspection.

If there are no corrections, the building inspector checks this box.

If there are corrections, the building inspectors checks this box.

This is the date the building official or inspector signs the memorandum.

This is for the signature of the inspector who performed the inspection.

Most notice forms have a telephone number so you can call the building department to set up the inspection.

Figure 6–3 Sample form of Inspection Record Memorandum.

The work must then be corrected and must not be covered up or concealed until the building official says it's okay to do so.

What Happens When The Permit Holder Does Not Call For An Inspection

If you do not call for a required inspection, you may be charged with a violation for not calling for the required inspection, working without the approval of the building official, or concealing work before an inspection. You may then have to go to court and pay a fine. The building inspector may also issue a stop work order. Also, you will not be able to obtain a certificate of occupancy. This means that you will not be able to use or occupy the structure. If you do so without a certificate of occupancy, you can be prosecuted and forced to pay fines. You must pass all of your required inspections in order to use the structure; therefore, you must call the building department to schedule those inspections.

If you want to change the plans during the building process, you must obtain permission first. You need to submit a request for a change order and have it reviewed by the building department. Make sure you get the inspector's permission in writing. You don't want this to become an issue during the final inspection. If you don't ask for permission, the building inspector can ticket you for deviating from the approved plans and can issue a stop work order.

Building officials frequently complain that permit holders delay calling for the final inspection. The permit holder then becomes desperate to obtain the final inspection before the sale of the finished building is due to take place. Unfortunately, that situation happens quite frequently.

In the Field

"I had a builder who hadn't called for a final inspection. He calls me at 3:00 p.m. on a Friday demanding I do the inspection right away because his buyer was at the real estate closing and the mortgage company wouldn't fund the purchase without a certificate of occupancy inspection. I already had an inspection scheduled and I told him I couldn't do it until Monday. I heard he had to pay for a motel and storage fees for the buyers because he waited until the last minute." —*John, Building Official* ■

PERMIT AND INSPECTION PROCESS

The accompanying chart shows the sequence of the permit and inspection process. One thing flows from the other. You can't have an inspection if you don't have a permit for it. You won't get an inspection if you don't call for one. You will not receive a certificate of occupancy until the work has passed the final inspection. Give yourself plenty of time when you estimate how long this process will take.

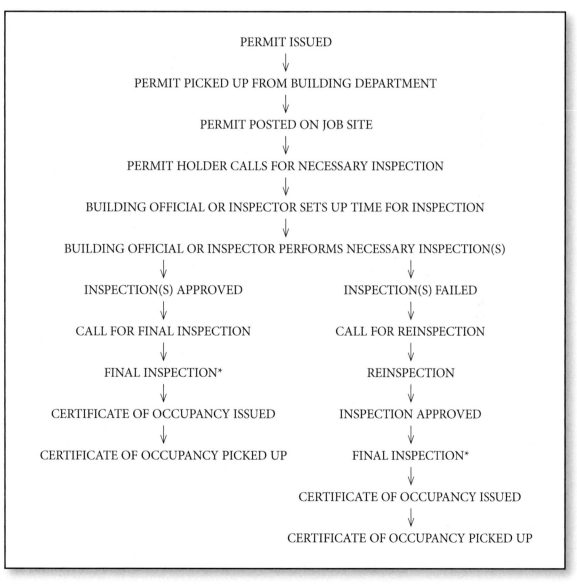

PERMIT ISSUED
↓
PERMIT PICKED UP FROM BUILDING DEPARTMENT
↓
PERMIT POSTED ON JOB SITE
↓
PERMIT HOLDER CALLS FOR NECESSARY INSPECTION
↓
BUILDING OFFICIAL OR INSPECTOR SETS UP TIME FOR INSPECTION
↓
BUILDING OFFICIAL OR INSPECTOR PERFORMS NECESSARY INSPECTION(S)
↓   ↓
INSPECTION(S) APPROVED   INSPECTION(S) FAILED
↓   ↓
CALL FOR FINAL INSPECTION   CALL FOR REINSPECTION
↓   ↓
FINAL INSPECTION*   REINSPECTION
↓   ↓
CERTIFICATE OF OCCUPANCY ISSUED   INSPECTION APPROVED
↓   ↓
CERTIFICATE OF OCCUPANCY PICKED UP   FINAL INSPECTION*
                                              ↓
                                  CERTIFICATE OF OCCUPANCY ISSUED
                                              ↓
                                  CERTIFICATE OF OCCUPANCY PICKED UP

* Note: In cases in which only one inspection is necessary, eliminate the final inspection step.

Figure 6–4 Permit and Inspection Process.

Required Inspections

Certain kinds of inspections are required by every building code. In order to know for sure which ones are required in your area, you must ask the building department or read the department's guidelines for inspections. Following are typical lists based on the 2006 International Residential Code and the International Building Code. The "Nature of Inspection" description is in the first column of each table. The time when the inspection must be performed is in the second column. The third column is the section number of the specific building code that requires the inspection. Remember that these are examples of required inspections from two model building codes. Because these codes are frequently amended by the local jurisdiction, you must check with the local building department for its schedule of inspections. Many building departments give you a list of the required inspections when you pick up your permit.

TABLE 6-1 Required Inspections, International Residential Code.

NATURE OF INSPECTION	WHEN PERFORMED	SECTION
Foundation inspection	After poles or piers are set or trenches or basement areas are excavated and any required forms erected and any required reinforcing steel is in place and supported prior to the placing of the concrete and must include excavation for thickened slabs intended for the support of bearing walls, partitions, structural supports or equipment, and special requirements for wood foundations	IRC R109.1.1
Plumbing, mechanical, gas, and electrical systems	Rough inspection must be made prior to covering or concealment, before fixtures or appliances are set or installed, and prior to framing inspection except back-filling of ground-source heat pump loop systems tested in accordance with Section M2105.1 prior to inspection is permitted	IRC R109.1.2
Lowest floor elevation	In areas prone to flooding pursuant to Table R301.2(1) upon placement of the lowest floor, including the basement, and prior to further vertical construction, must submit documentation, prepared by a registered design professional, of the elevation of the lowest floor, including the basement required by IRC R324, to building official	IRC R109.1.3
Frame and masonry inspection	After roof, masonry, all framing, firestopping, draftstopping, and bracing are in place and after the plumbing, mechanical, and electrical rough inspections are approved	IRC R109.1.4
Other inspections	When needed to ascertain compliance with provisions of the code and other laws	IRC R109.1.5
Fire-resistant-rated construction inspection	Where fire-resistance-rated construction is required between dwelling units or due to location on the property, inspection	

continues

NATURE OF INSPECTION	WHEN PERFORMED	SECTION
	must be done after all lathing and/or wallboard is in place, but before any plaster is applied or before wallboard joints and fasteners are taped and finished	IRC R109.1.5.1
Reinforced masonry, insulating concrete form, and conventionally formed concrete walls	Where located in Seismic Design Categories D_0 or D_1, D_2 and E, inspections must be conducted after plumbing, mechanical, and electrical systems are embedded within the walls and reinforcing steel are in place and prior to placement of grout or concrete. Must verify the correct size, location, spacing, and lapping of reinforcement. Must verify that the location of grout cleanouts and size of grout spaces comply with the requirements of the code	IRC R109.1.5.2
Final inspection	After all work required by the building permit is completed	IRC R109.1.6

TABLE 6-1   Required Inspections, International Residential Code.   *continued*

Figure 6–5   Building inspector looking at unfinished stucco wall.

Figure 6–6   Building inspector looking at framework.

Figure 6–7 Building inspector looking at plans and electrical panel.

TABLE 6-2 Required Inspections, International Building Code.

NATURE OF INSPECTION	WHEN PERFORMED	SECTION
Footing and foundation inspection	After excavations for footings are complete and any required reinforcing steel is in place; required forms must be in place prior to inspection and materials for foundation must be on the job except where concrete is ready mixed in accordance with ASTM 94	IBC 109.3.1
Concrete slab and under-floor inspection	After in-slab or under-floor reinforcing steel and building service equipment, conduit, piping accessories, and other ancillary equipment items are in place, but before any concrete is placed or floor sheathing installed, including the subfloor	IBC 109.3.2
Lowest floor elevation	In flood hazard areas upon placement of the lowest floor, including the basement, and prior to further vertical construction, must present elevation certification required by IBC 1612.5 to building official	IBC 109.3.3
Frame inspection	After roof deck or sheathing, all framing, fireblocking, and bracing are in place and pipes, chimneys, and vents to be concealed are complete, and the rough electrical, plumbing, heating wires, pipes, and ducts are approved	IBC 109.3.4
Lath and gypsum board inspection	After lathing and gypsum board, interior and exterior, is in place but before any plastering is applied or gypsum board joints and fasteners are taped and finished except for gypsum board not part of a fire-resistance-rated assembly or shear assembly	IBC 109.3.5
Fire-resistant penetrations	Protection of joints and penetration in fire-resistance-rated assemblies not to be concealed from view until inspection and approval	IBC 109.3.6
Energy efficiency inspections	Must be made to determine compliance with Chapter 13 and must include inspections for envelope insulation R and U	

continues

TABLE 6-2 Required Inspections, International Building Code. *continued*

NATURE OF INSPECTION	WHEN PERFORMED	SECTION
	values, fenestration U value, duct system R value, and HVAC and water-heating equipment efficiency	IBC 109.3.7
Other inspections	When needed to ascertain compliance with provisions of the code and other laws	IBC 109.3.8
Special inspections	As required by Chapter 17, Structural Tests and Special Inspections	IBC 109.3.9
Final inspection	After all work required by the building permit is completed	IBC 109.3.10

Builders run into trouble when they move ahead in the process without getting an inspection first. Common problems are framing the structure before having the necessary footing and foundation inspection and covering up work in the wall before calling for the inspection.

In the Field

"Builders in our village are always getting into trouble because they begin framing the structure before calling for the footing and foundation inspection. They pay hundreds of dollars in fines. I just don't get it because we are very good about getting out to the job site and checking the work so we don't delay the construction process. Maybe they get away with it in other towns and aren't penalized. Not in our village." —*Tim, Deputy Building Official* ∎

To help you make sure you haven't overlooked anything regarding inspections, use this checklist as a reminder.

JUST ✔ CHECKING

Inspections

☐ What inspections are required to be performed based on the nature of my approved permit?
    ☐ Footing
    ☐ Foundation

continues

JUST ✔ CHECKING

Inspections (continued)

- ☐ Backfill
- ☐ Spot survey
- ☐ Sewer and water
- ☐ Electrical service
- ☐ Electrical rough
- ☐ Mechanical
- ☐ Plumbing rough-in
- ☐ Plumbing underground
- ☐ Framing
- ☐ Insulation
- ☐ Basement, garage, slab floor
- ☐ Sidewalk
- ☐ Topographical survey/flood plain
- ☐ Fire-resistance-rated construction
- ☐ Reinforced masonry, insulating concrete form, and conventionally formed concrete wall
- ☐ Final building and plumbing
- ☐ Final electrical
- ☐ Final engineering
- ☐ Special _____
- ☐ Other_____
- ☐ Is it time to call for an inspection?
- ☐ Who is responsible for calling for the inspection?
- ☐ Have I notified the building official that the work is ready to be inspected?
- ☐ Did I document my call for an inspection?
- ☐ What documentation exists to prove an inspection took place?
- ☐ Has the work passed each necessary inspection?
- ☐ Is there written approval for the inspection?
- ☐ If the work has not passed the inspection, what corrections must be made in order to obtain approval?
- ☐ Have I notified the building official that the corrected work is ready to be inspected?
- ☐ Has the work passed the subsequent inspection?

Inspections (continued)

- ☐ Has any of the work been covered up or made inaccessible prior to approval?
- ☐ Do I have to uncover the concealed work in order that an inspection can take place?
- ☐ Should I ask the building official for a preliminary inspection of a building, structure, or site to take place before the permit is issued?
- ☐ Are there any other inspections that are required to make sure there is compliance with the provisions of the building code and other laws that are enforced by the Department of Building Safety?
- ☐ What special inspections need to be performed?
- ☐ Is the work ready for a final inspection?
- ☐ Has the work passed any of the required inspections?
    - ☐ Building
    - ☐ Electrical
    - ☐ Mechanical
    - ☐ Plumbing
    - ☐ Concrete
    - ☐ Engineering
    - ☐ Water
    - ☐ Sewer
    - ☐ Fire
    - ☐ Health
    - ☐ Landscaping
    - ☐ Energy
    - ☐ Medical Gas
    - ☐ Zoning
    - ☐ Historical
    - ☐ Special Inspections
    - ☐ Site Plan
    - ☐ Stormwater Management
    - ☐ Special _____
    - ☐ Other _____
- ☐ Can I get a certificate of occupancy?

Once you have passed all of the required inspections, you will be ready and eligible to obtain a certificate of occupancy.

AVOIDING TROUBLE

- Check to find out what inspections you need.
- Call for an inspection in a timely fashion.
- Complete corrections and call for a reinspection as soon as possible.
- Don't cover up work that must be inspected.
- Get permission for changes in writing.
- Don't wait until the last minute to get a final inspection.

CONCLUSION

Once you obtain your permit, call for all required inspections in a timely manner. Know what these inspections are by checking the building code and by asking the building department for a list. If a problem is found during an inspection, correct it in a reasonable amount of time. Don't move ahead with construction if you still need a required inspection. Give yourself plenty of time to schedule an appointment with the building department so you don't find yourself explaining why you can't get a certificate of occupancy.

CHAPTER 7

STOP WORK ORDERS

INTRODUCTION

Homeowners and contractors dread a stop work order more than anything else. A building official has the authority to issue a stop work order regarding any work regulated by the building code that is performed in a manner either contrary to the provisions of the code or that is dangerous or unsafe. This means that if you perform work without a permit or contrary to the approved plans, the building official can make you stop construction.

ISSUANCE OF A STOP WORK ORDER

Stop work orders are most often issued when an inspector comes upon work being done without a permit. This is usually by happenstance.

KEY CONCEPTS

stop work order—an order from a building official that work must cease at a building site because of a violation of the building code or because the work being performed is unsafe.

CONSTRUCTION PERMIT

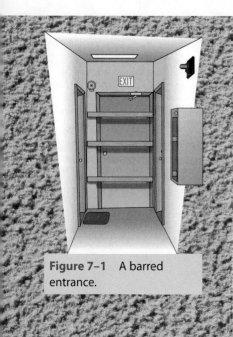

Figure 7–1 A barred entrance.

In the Field

"I was doing my annual inspection of an apartment building when I noticed that there was construction going on without a permit. The contractor had closed off the exit to the stairwell for the second floor by nailing a board across the door. He was replacing the staircase in the stairwell. Doing this without a permit was bad enough, but it also meant that if there had been a fire in the building, residents on the second floor would have only had one exit to use to get out of the building because the second exit required by the fire code was completely blocked. I issued a stop work order and ticketed the owner. He ended up paying a few hundred dollars in fines and couldn't finish the work until he applied for permits. He also had to set up a special fire watch because the closed off stairwell couldn't be used, which I'm sure cost him a lot of money." —*Kathy, Building Inspector* ∎

Power Of The Building Official To Issue Stop Work Orders

Building officials seldom need to take violators to court in order to gain compliance because of their power to issue a stop work order. The reason is very simple. If a contractor wishes to continue a project, she must obey the building official's order and correct any

In the Field

"I had a builder completely change the dimensions on a building from the approved plans. They were so far off that the foundation encroached into the side-yard setback required by the zoning code. I issued a stop work order and the builder had to wait while revised plans were drawn and approved. He had to remove the foundation that had already been poured. The whole thing cost him about $5000." —*Eric, Building Official* ∎

problems. Although the violator can always appeal the decision of the building official, the resulting delay in construction makes it undesirable, especially when there is no basis on which to challenge the decision of the building official. A stop work order can be for the entire job site or for a portion of the work, depending on the facts of the situation. Always get approval from the building department before deviating from the approved plans. You'll spare yourself a stop work order.

Contents Of A Stop Work Order

This is what a stop work order looks like:

DEPARTMENT OF BUILDING SAFETY
STOP WORK ORDER
1181 Paulina Dr., Woodridge, Illinois
Site Address

The address listed above is in violation of the Village of Woodridge Code 2006 IBC 105.1 as amended and adopted by reference in Section 8-1D-1 of the Village of Woodridge Code, Working without a Permit. All work shall be stopped and shall not continue until approval is granted from the Building Official and all necessary permits are obtained.
For information, contact the Department of Building Safety at 630-555-4750.

WARNING
This notice is to be removed only by the authority of the Building Official.
Date Posted: 7/18/XX   By: John Black

Building Official
PENALTY IF REMOVED WITHOUT AUTHORIZATION.
To file an appeal to this order, send notice in writing to the Village of Woodridge, Board of Appeals, One Plaza Drive, Woodridge, IL 60517.

Address of the job site.

Section of the code that has been violated.

Order of the building official to stop work and not continue.

Conditions for the stop work order to be lifted.

Notice that if the stop work order is removed, a violation may be given to the responsible party.

Date of the order.

Signature of the building official having authority.

Appeal rights if there is a disagreement as to whether the order should have been issued.

Figure 7–2   Sample Stop Work Order Form.

Stop Work Order Procedure

It is a serious matter to get served with a stop work order. You should not ignore it or you'll get in even more trouble. Once a stop work order has been issued, you may not perform any other work unless you have been directed to do so in order to remove the violation or unsafe condition. If you continue to perform the work, you and anyone else served with the stop work order can be ticketed for violating the stop work order and maybe even arrested, depending on the local law. You will also earn the everlasting distrust of the building inspector.

Figure 7–3 If you continue construction after being served with a stop work order, you and anyone else served can be ticketed for violation and, depending on the local law, perhaps arrested.

In the Field

"I was doing my annual fire inspections when I noticed that some people had opened up a spray painting business without a permit and without the necessary safeguards. I issued a stop work order and told them they needed to come into the village to apply for an operational permit and that they had to follow the safety regulations. Well, I went back later in the afternoon and they had continued to do the work. I called the police department and had them arrested for violating my stop work order. They had to go to the police station, post bail, and go to court to pay their fines." —*Trevor, Fire Inspector* ■

This flowchart explains the procedure involved in issuing a stop work order:

INSPECTION BY BUILDING OFFICIAL OR INSPECTOR
↓
BUILDING OFFICIAL OR INSPECTOR FINDS REASON FOR STOP WORK ORDER
↓
BUILDING OFFICIAL ISSUES STOP WORK ORDER
↓

Figure 7–4 Issuing a Stop Work Order.

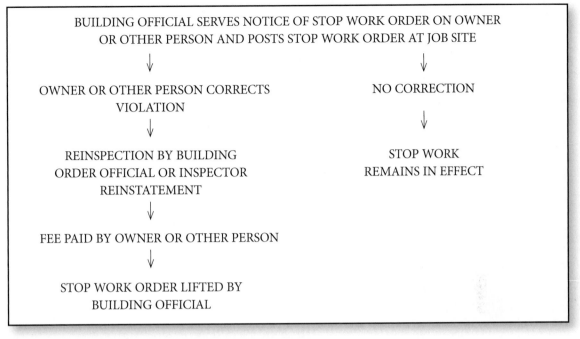

BUILDING OFFICIAL SERVES NOTICE OF STOP WORK ORDER ON OWNER
OR OTHER PERSON AND POSTS STOP WORK ORDER AT JOB SITE

↓   ↓

OWNER OR OTHER PERSON CORRECTS   NO CORRECTION
VIOLATION

↓   ↓

REINSPECTION BY BUILDING   STOP WORK
ORDER OFFICIAL OR INSPECTOR   REMAINS IN EFFECT
REINSTATEMENT

↓

FEE PAID BY OWNER OR OTHER PERSON

↓

STOP WORK ORDER LIFTED BY
BUILDING OFFICIAL

Figure 7–4   *continued*

If you are served with a stop work order, you will usually have
to pay a reinstatement fee. Some building departments charge as
much as $500 to lift the stop work order. If you have been working
without a permit, you will need to obtain a permit for the work you
have already done and any work remaining. Building departments
increase their fees for people who don't obtain a permit for work,
often doubling the fees.

Figure 7–5   Posted
Stop Work Order.

JUST ✔ CHECKING

Stop Work Orders

☐ What is the nature of the work being performed that is contrary to the building code?

☐ Is the work being performed dangerous or unsafe?

☐ Has the stop work order been reduced to writing?

☐ Does the stop work order contain the reason for the order and the conditions under which the cited work will be permitted to resume?

☐ Who has been served with the stop work order?
  ☐ Owner?
  ☐ Owner's agent?
  ☐ Person doing the work?

☐ How was service made?
  ☐ Posted in a conspicuous place in or about the structure affected by the stop work order
  ☐ Delivered personally
  ☐ Sent by certified mail addressed to the last known address
  ☐ Sent by first-class mail addressed to the last known address
  ☐ Private delivery service

☐ Is it possible to get the stop work order lifted?

☐ Should the stop work order be appealed?

☐ Has the fee to lift the stop work order been paid?

☐ Can other work continue during the stop work order?

Just because you have applied for a permit doesn't mean you can begin doing the work. Until you have the permit posted on the job site, any work you do is illegal and subject to a stop work order.

In the Field

"I had a builder who did the right thing by applying for a building permit to renovate a storefront, but then let his workers begin demolition and construction. I had to issue a stop work order and his project ended up being delayed, and he had to pay a fee to lift the stop work order, all because he didn't want to wait for approval." —*Eric, Building Official* ■

AVOIDING TROUBLE

- Never do work without the necessary permits.
- Always follow the approved plans.
- Always seek the approval of the building department if you need to change the plans.
- Always make sure the work is being performed in a safe manner.
- When in doubt, call the building inspector before proceeding with the work in question.

CONCLUSION

People who are served with a stop work order usually get one because they have tried to take a short cut. Some builders do work without a permit, perform work beyond the scope of the permit, or create unsafe conditions. By getting the necessary permits, following the approved plans, and maintaining a safe job site, you will not have to worry about receiving a stop work order.

CHAPTER 8

CERTIFICATES OF OCCUPANCY

INTRODUCTION

Most building codes require that a building or structure have a certificate of occupancy before the building or structure is used or occupied. This is to make sure the building or structure is safe before persons are allowed to use the building. A certificate of occupancy is normally issued after the building has passed the final inspection after a permit.

WHEN A CERTIFICATE OF OCCUPANCY IS NECESSARY

In order to use or occupy a building, new construction requires a certificate of occupancy. For example, the 2006 International Residential Code states as follows:

> R110.1 Use and occupancy. No building or structure shall be used or occupied, and no change in the existing occupancy classification of a building or structure or portion thereof shall be made until the building official has issued a certificate of occupancy therefore as provided herein.

A certificate of occupancy is not usually needed if the work performed is exempt from the requirement for a permit. Accessory buildings or structures don't usually require a certificate of occupancy.

In addition to new construction, the existing occupancy classification of a building or structure may not be changed until the building official issues a certificate of occupancy. One reason for this is because the type of fire protection necessary for the new use may not be sufficient if the use is changed. Therefore, before new occupants are able to use the building, there must be a new certificate of occupancy if the occupancy classification is different. For example, if the use is changed from a business office to a day-care center, that is a change of use. When in doubt, contact the local building and/or fire department.

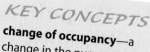

KEY CONCEPTS

change of occupancy—a change in the purpose or level of activity within a building that involves a change in application of the requirements of the code.

International Fire Code

Figure 8–1 The type of fire protection necessary for building use may not be sufficient if the use is changed.

JUST ✔ CHECKING

CHECKLIST—Certificates Of Occupancy

☐ Does the building or structure have a certificate of occupancy?
☐ Was the building constructed before certificates of occupancy were issued by the local jurisdiction?
☐ Have all the necessary inspections been performed and passed so the certificate of occupancy can be issued?
☐ Has the building's or structure's classification changed?
☐ Is there a certificate of occupancy for the classification change?

continues

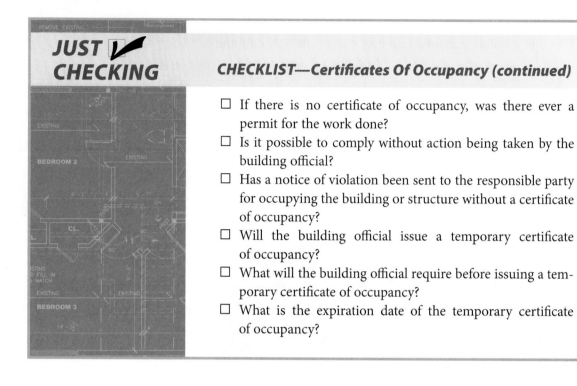

CHECKLIST—Certificates Of Occupancy (continued)

☐ If there is no certificate of occupancy, was there ever a permit for the work done?

☐ Is it possible to comply without action being taken by the building official?

☐ Has a notice of violation been sent to the responsible party for occupying the building or structure without a certificate of occupancy?

☐ Will the building official issue a temporary certificate of occupancy?

☐ What will the building official require before issuing a temporary certificate of occupancy?

☐ What is the expiration date of the temporary certificate of occupancy?

Unintended Consequences—Failing To Obtain A Certificate Of Occupancy

It is the responsibility of the homeowner or contractor to call for the final inspection so that a certificate of occupancy can be issued. Building officials are too busy to remind permit holders that they need to call for a final inspection so that a certificate of occupancy can be issued. Many inspectors have drawers full of permits that have not been "finaled out," that is, received a final inspection.

However, if it comes to the building official's attention that a building or structure has been occupied unlawfully, there will be unpleasant consequences. The building official may order that the occupants immediately vacate the building. He may also issue a notice of violation and bring the violator into court.

The failure to get a certificate of occupancy for a residence can have a domino effect on a new buyer. Homebuyers often sell

In the Field

"I had a commercial building where the occupants got ahead of the process and moved things in overnight, thinking they would be able to get the certificate of occupancy after the fact. Unfortunately for them, they had a new fire alarm and fire suppression system that hadn't passed a final inspection. They had really bad luck because there were two pull stations, one of which was dead, and it took quite a bit of time to figure out which one was the problem. We made them move all of their equipment and furniture out of the building because they didn't have a certificate of occupancy and they were delayed in opening up their business." —*Eric, Building Official* ∎

their home the same day as they buy the new one. They often move all of their belongings out of the first residence before they've bought the second one. Don't rush the building process; give yourself plenty of time before moving in so you can get a certificate of occupancy.

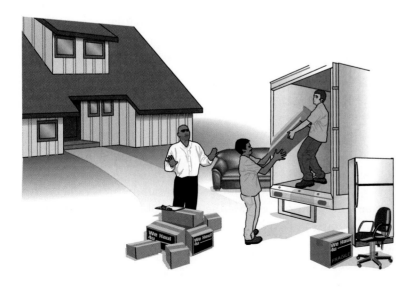

Figure 8–2 If there is no certificate of occupancy, the building inspector will stop new residents from moving in.

In the Field

"One time there was a builder who tried to rush me into giving him a certificate of occupancy because he had set a closing date and the buyers wanted to move in. Some water had gotten into the building because of a bad storm, and instead of leaving the walls open like he was supposed to so I could see if there was a problem with the electrical work and the insulation from the water, he went ahead and drywalled right over the area I needed to see. I told him I wouldn't give approval until he removed the drywall. He must have called me every name in the book before he told his buyers, who had all of their possessions loaded on the moving van, that they couldn't move in. They had sold their house that morning and had no place to go. I felt sorry for them, but they chose that contractor. If they had asked around, they would have found out a lot of bad things about the way he does business." —*Tim, Building Official* ■

Problems In Existing Building

If you buy property without doing your due diligence, you may find yourself with a building or structure that lacks a certificate of occupancy for the entire building or for a part of it. For example, homeowners often finish their basements or construct decks without getting a permit. The house is then sold without disclosing that the property is in violation of the code. Somehow the building official becomes aware of the problem, usually when he is inspecting the property for another reason. The building official can't go after the original owner because too much time has gone by to charge the violator with not getting a permit. When this happens, the building official cannot file a violation for the permit violation because the statute of limitations may have run out. Once the time has passed, no charges can be brought against the violator for the permit violation, no matter how outrageous it is.

Figure 8–3   Older home with newly-built porch.

However, every day the property is occupied or used without the required certificate of occupancy is a new violation. Therefore, the building official can bring a charge against the new owner for violating the certificate of occupancy ordinance if the evidence is sufficient.

Missing Certificates Of Occupancy

If you are an owner who didn't know about the violation, the best thing to do is to work with the building official in order to obtain the certificate of occupancy. The situation may be simple; work has been done with a permit but the permit holder has not called for a final inspection, and the file remains open. The building official may send you an informal letter seeking to resolve the matter. If you obtain the final inspection, the inspector can then issue a certificate of occupancy. But, if you don't allow the final inspection, the building official may take enforcement action. The building official's primary concern is to make sure the building has no serious health or safety issues and that there have been no alterations done without permits. Some jurisdictions handle these matters informally by requesting an inspection and issuing a letter of compliance so there is a record that the building has been checked for health or safety issues as of a particular date. But, if the owner won't cooperate, consequences can be severe.

In the Field

"We were catching up on our permits where no one had called for a final inspection. We sent letters to the residents asking them to let us complete our inspection and issue a certificate of compliance so we could close our files. One couple refused to let us inspect a pretty large addition, telling us to talk to the previous owner. We warned them that we'd have to issue a notice of violation for occupying the structure without a certificate of occupancy but they just ignored us. We had to get an administrative search warrant to do our inspection. During our inspection, we found such bad electrical and mechanical problems that we ordered them not to occupy the space. They refused to obey my order so we took them to court. They ended up having to spend thousands of dollars to make the necessary repairs, pay $1,200 in fines, and I'm sure they paid their attorney a lot of money. It was all so unnecessary. We would have worked with them if they had shown the least inclination to cooperate." —*Roger, Building Official* ■

If you are an innocent purchaser of property with code violations, you may want to talk to a lawyer about filing a lawsuit against the seller for fraud or even rescission of the purchase. You can subpoena the building official to testify as your witness.

In the Field

"I really felt sorry for these people. They bought this house thinking they could use it for their family. They had five children. We have an ordinance against over-occupancy to prevent tragedies. People die in fires because they sleep in uninhabitable spaces and can't escape from a basement, for example, that has no escape window. It was a very small house only suitable for three people. The realtor had told them they could use the basement rooms for bedrooms, which was totally untrue. We suggested they hire a lawyer to help them since they couldn't live in the house with seven people because it was a violation of our ordinance. I even offered to testify for them against the seller." —*Juan, Building Inspector* ■

TEMPORARY OCCUPANCY PERMITS

Some jurisdictions will allow occupants to move into a building even though certain items need to be completed, such as landscaping or a driveway. You will usually have to post bond or place money in escrow so the local jurisdiction is protected if you fail to fulfill the condition of the temporary certificate of occupancy.

You still need to get a regular certificate of occupancy even if you have a temporary one. Once the condition is fulfilled, you need to call the building official for the final inspection. If you don't call and do this, you may find yourself being charged with occupying a building without a certificate of occupancy. Most temporary certificates have an expiration, date and you must complete the work prior to the expiration date and obtain your certificate of occupancy as well. Also, if you don't fulfill the conditions of the temporary certificate, the building official may try to revoke it.

Temporary certificates of occupancy are disliked by building officials because of the headaches they can create. Temporary occupancy certificates are a huge problem for local jurisdictions. Once the structure is occupied, judges are reluctant to evict the occupants. Therefore, building officials issue temporary occupancy certificates with great restraint. Don't count on getting a temporary occupancy permit so you can move in before work is completed.

In the Field

"I really felt sorry for Mr. and Mrs. C. They obtained a temporary occupancy permit because the grading had not yet been completed for their newly constructed home. It was winter time and the work couldn't be done. The temporary occupancy permit expired six months after it had been issued. The builder never got around to submitting the proper plan to complete the grading work. Two years later the village was conducting a review of its temporary occupancy permits that had never been converted to certificates of occupancy and discovered that the grading work had never been completed. Mr. and Mrs. C had

continues

In the Field (continued)

money in escrow to complete the work with the title company, but the title company wouldn't release the cash without a certificate of occupancy. Mr. and Mrs. C didn't have the money to hire someone new to do the grading, and we couldn't prosecute the original builder because he no longer owned the property. Mr. and Mrs. C were the ones living in it. They were taken to court for occupying a building without a certificate of occupancy. They ended up threatening to sue the builder. He did the work but he wanted to do it on the cheap, so it took much longer to get our approval because he kept messing up the job. Once the work was finally done, I asked that the charges against Mr. and Mrs. C. be dropped." —*Brian, Building Inspector*

VIOLATIONS

If you get caught without a certificate of occupancy, the building official will usually send you a notice of violation. If you comply, that will be the end of the matter. But, if you ignore the notice, the local jurisdiction can issue a ticket or even sue you in court to stop you from using the property.

This chart shows what the procedure is if you get caught without a certificate of occupancy.

BUILDING OFFICIAL DISCOVERS EVIDENCE OF OCCUPANCY
OF A STRUCTURE OR BUILDING WITHOUT A CERTIFICATE OF OCCUPANCY

↓

BUILDING OFFICIAL SENDS INFORMAL LETTER TO RESPONSIBLE PARTY

OR

BUILDING OFFICIAL SENDS NOTICE OF VIOLATION AND PREPARES
PROOF OF SERVICE

↓

Figure 8–4 Enforcement of certificate of occupancy violations.

BUILDING OFFICIAL REINSPECTS PROPERTY AFTER
COMPLIANCE DATE

↓

CASE CLOSED BECAUSE OF COMPLIANCE

OR

PREPARATION OF COMPLAINT BY BUILDING OFFICIAL OR LEGAL
COUNSEL AGAINST OWNER

↓

COMPLAINT SUBMITTED TO COURT OR TO ADMINISTRATIVE
HEARING BODY

↓

DEFENDANT IS SERVED WITH COMPLAINT ACCORDING TO RULES OF LOCAL
JURISDICTION

↓

DEFENDANT GOES TO COURT OR ADMINISTRATIVE HEARING

Figure 8–4 *continued*

AVOIDING TROUBLE

- Never occupy or use a building or structure without a certificate of occupancy.
- Call the building official or inspector when you are ready for a final inspection.
- If you are given a temporary certificate of occupancy, complete all the conditions in a timely fashion and call the inspector when you are ready for a final inspection.
- When buying property, check with the local building department to see if the building or structure has a certificate of occupancy and if there is anything you need to do before moving into it.
- If you are moving into property and are going to change the use of the property, call your local building department and fire department to see what you need to do to get a new certificate of occupancy.

CONCLUSION

Whenever you are planning to occupy a building, check with the local building department to make sure you can legally occupy the space. If you are a builder, make sure all the necessary inspections are performed so you can obtain a certificate of occupancy before the building is used. If you follow the rules of the local jurisdiction, you should have no problem obtaining a certificate of occupancy on time.

CHAPTER 9

GETTING IN TROUBLE

INTRODUCTION

Sometimes despite the best of intentions, you may find yourself in violation of some provision of the building code. Usually the owner of the property is responsible, along with anyone who performed the unlawful act that violates the code. Therefore, you may find yourself having to defend what a contractor did on your behalf, or, if you are a contractor, what one of your employees did.

WHAT LEADS TO PROBLEMS

Prior to prosecuting a violator for an unlawful act, the building inspector usually seeks to obtain compliance by talking to the responsible party so the problem can be fixed. But, if the party ignores the building inspector, she may have to serve a notice of violation or order on the person responsible for the unlawful act. Too often this happens because the person doesn't know what they're doing. You are expected to have the expertise to build according to the code whether or not you actually do. It's not a defense that you didn't know any better.

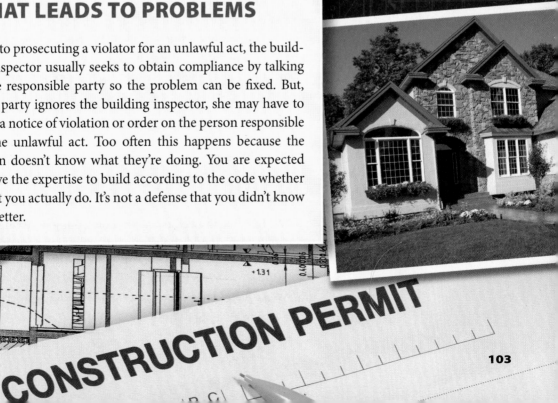

CONSTRUCTION PERMIT

Figure 9–1 You are expected to have the expertise to build according to the code whether or not you actually do. Work with building officials to better understand it.

In the Field

"I had a situation where the stairs on a deck didn't meet the requirements of the code. I went out to the job site with my code book and showed the builder the code. I asked him, 'Do you agree with it?' He said 'Yes, but do you expect me to know that?' This was a guy who was licensed, and he got belligerent with me because I expected him to know the code and build accordingly. I can't believe people were paying this guy to do the work. He kept arguing that the code was stupid, so I finally had to issue a notice of violation." —*Eric, Building Official* ■

NOTICE OF VIOLATION

KEY CONCEPTS

notice of violation—an official document issued by a building official to a person responsible for a violation of the building code demanding correction of the violation within a specific time frame.

It does not make sense to ignore a notice of violation from a building inspector. Since most inspectors just want compliance, you should at least acknowledge the notice of violation and get in touch with the inspector to tell him how you plan to comply with his order. Why subject yourself to delays and fines if you don't have to?

Most codes authorize the building official to serve a notice of violation or an order on the responsible person. Usually that notice will be in writing. The notice will contain a time for compliance. If you pay attention to the notice and fix the problem, your trouble ends there.

In the Field

"We building inspectors are just doing our jobs, trying to protect people from shoddy construction by making sure minimum standards are met. We're not out to make people miserable or add to their costs. But, we can't help people if they won't talk to us. I just don't understand why they won't pick up the phone and call us after we give them a notice of violation." —*Rob, Building Official* ■

Following is an example of what a notice of violation looks like.

October 3, 20XX

Mrs. Karyn Byrne
6560 Hollywood Blvd.
Hinsdale, IL 60521

Re: 6560 Hollywood Blvd., Hinsdale, Illinois
Property Index Number: 04-0004-004-00

Dear Mrs. Byrne:

An inspection of your property at 6560 Hollywood Blvd., Hinsdale, Illinois, on October 2, 20XX shows the following violations of the code of ordinances of the Village of Hinsdale:

Using a Structure without a Certificate of Occupancy in violation of 2006 IRC R110.1 as amended and adopted by reference in Section 9-3-1 of the Village of Hinsdale Code—an addition is being used without a certificate of occupancy. A permit was obtained on June 11, 20XX, for the addition, but a final inspection was never performed.

You must discontinue this illegal activity and abate the violation. You must cease use of the addition until a certificate of occupancy is obtained. You need to arrange for a final inspection by October 20, 20XX, and pay all necessary fees for a reinstatement or extension of the prior permit. A certificate of occupancy must be obtained by the close of business on November 22, 20XX, or a complaint will be filed against you in a court of local jurisdiction.

Please feel free to contact me to discuss this matter further.

Very truly yours,

Robert McGinnis

Building Official

Callout labels:
- Name of responsible party.
- Address of site of violation.
- Name of violation.
- Section number of code violated.
- Date of inspection.
- Order of abatement.
- Deadline for obtaining inspection.
- Deadline for obtaining certificate of occupancy.
- Name of the building official to contact.

Figure 9–2   Example notice of violation to responsible party for no certificate of occupancy.

Response To A Notice Of Violation

Whatever you do, be truthful with the inspector. Inspectors, like everyone else, like it when people are candid with them. They are more likely to help you or to give you a break and not write you a ticket when you are honest. If you try to deceive the inspector, your credibility will be destroyed and anything you say can be used against you in court. If you are a contractor, don't be surprised if your reputation for deception spreads among the community of building inspectors. The grapevine is powerful.

In the Field

"I noticed that an addition had been added to a residence when I was in the neighborhood checking a neighbor's project. I checked the village's records and found that no permit had ever been issued for the structure and that it violated the setback rules of the local zoning code. The homeowner was given a notice of violation. She contended that she had bought the house that way 30 years ago. Under those circumstances the addition would have been grandfathered in and allowed. However, I pulled out aerial photographs that had been taken by the county from 10 years earlier and discovered that the addition did not exist at that time. We ended up filing a lawsuit against her and she was forced to tear down the addition. I'm sure she spent thousands of dollars in legal fees." —*John, Building Official* ■

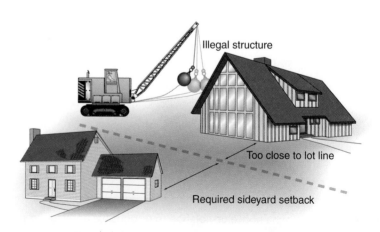

Illegal structure

Too close to lot line

Required sideyard setback

Figure 9–3 Additions that violate zoning laws could be subject to the wrecking ball.

If you do not agree with the decision of the building official, you may appeal her decision. This procedure will be discussed in Chapter 10, Appealing the Decision of the Building Official.

JUST ✔ CHECKING

CHECKLIST—Notice Of Violation

☐ Does the violation notice list a legitimate problem on the job site?

☐ Is it possible to comply with the notice of violation?

☐ How long will it take to comply?

☐ If it will take longer than the due date in the notice of violation, has the building official been contacted to seek an extension for compliance?

☐ Will the building official grant an extension of time for compliance?

☐ Should a notice of appeal of the building official's order be filed?

FILING CHARGES AGAINST THE RESPONSIBLE PARTY

If a person does not comply with the order of the building official, a charging document for the court or administrative body will be prepared by the building official, someone she designates, or the attorney for the jurisdiction. The charging document or ticket describes the details of the offense because a person is entitled to know what the charges are so he or she can prepare a defense.

This chart shows the procedure followed by the building official during an enforcement action.

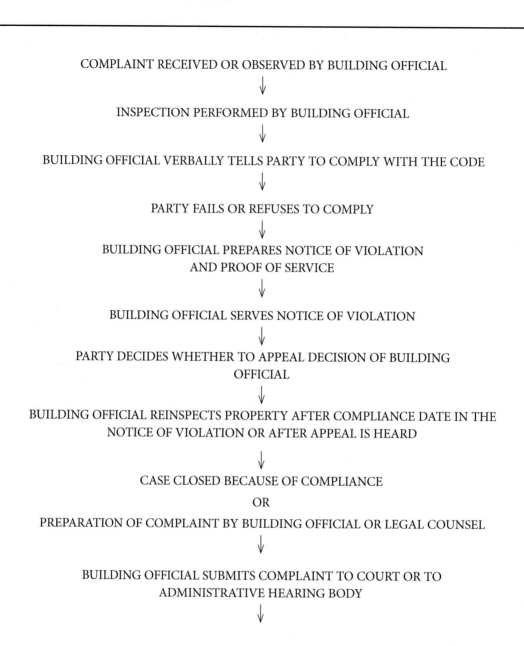

COMPLAINT RECEIVED OR OBSERVED BY BUILDING OFFICIAL

↓

INSPECTION PERFORMED BY BUILDING OFFICIAL

↓

BUILDING OFFICIAL VERBALLY TELLS PARTY TO COMPLY WITH THE CODE

↓

PARTY FAILS OR REFUSES TO COMPLY

↓

BUILDING OFFICIAL PREPARES NOTICE OF VIOLATION
AND PROOF OF SERVICE

↓

BUILDING OFFICIAL SERVES NOTICE OF VIOLATION

↓

PARTY DECIDES WHETHER TO APPEAL DECISION OF BUILDING
OFFICIAL

↓

BUILDING OFFICIAL REINSPECTS PROPERTY AFTER COMPLIANCE DATE IN THE
NOTICE OF VIOLATION OR AFTER APPEAL IS HEARD

↓

CASE CLOSED BECAUSE OF COMPLIANCE

OR

PREPARATION OF COMPLAINT BY BUILDING OFFICIAL OR LEGAL COUNSEL

↓

BUILDING OFFICIAL SUBMITS COMPLAINT TO COURT OR TO
ADMINISTRATIVE HEARING BODY

↓

Figure 9–4 Basic enforcement procedure for an unlawful act.

DEFENDANT IS SERVED WITH COMPLAINT ACCORDING TO RULES
OF LOCAL JURISDICTION

DEFENDANT GOES TO COURT OR ADMINISTRATIVE HEARING

DEFENDANT PLEADS GUILTY AND RECEIVES A PENALTY

OR

DEFENDANT PLEADS NOT GUILTY, GOES TO TRIAL OR HEARING,
IS FOUND GUILTY, AND RECEIVES A PENALTY

OR

DEFENDANT PLEADS NOT GUILTY, GOES TO TRIAL OR HEARING,
IS FOUND NOT GUILTY, CASE CLOSED

OR

CASE IS DISMISSED ON MOTION OF THE PROSECUTION OR COURT,
CASE CLOSED

Figure 9–4   *continued*

The Charging Document

Following is an example of a charging document for the notice
of violation shown earlier along with an explanation of its parts.
Every charging document should contain the responsible party's
name, the date of the offense, the name and section number of
the violation, a description of the violation, and the signature of
the complainant to satisfy the requirements that the defendant
receive due process.

STATE OF ILLINOIS

COUNTY OF DU PAGE

VILLAGE OF HINSDALE

v.

NAME: KARYN BYRNE

ADDRESS: 6560 Hollywood Blvd.

CITY: Hinsdale, Illinois 60521

State in which offense occurred.

County in which offense occurred.

Name of the local jurisdiction bringing the charge.

Name of the responsible party.

Address where the defendant can be served with the charge.

Date and time of the offense.

Name of the offense.

Section number of the offense.

The undersigned says that on or about October 6, 20XX at or about 4:00 p.m., the Defendant did unlawfully commit the offense of Using a Structure without a Certificate of Occupancy in violation of 2006 IRC R110.1 as amended and adopted by reference in Section 9-3-1 of the Village of Hinsdale Code, in that said Defendant, the owner of 6560 Hollywood Blvd., Hinsdale, Illinois, used a structure, being an addition that is part of a single-family residence on the premises, without a certificate of occupancy in that the structure was being used and had various items stored in it.

Robert McGinnis

Complainant

Sworn to and Subscribed Before Me
This 23rd Day of November, 20XX

Notary Public

The defendant's relationship to the property.

Description of the location of the violation.

Description of the violation.

Signature of the complaining witness.

Signature of a notary public.

Figure 9–5 Example form for occupying a structure without a certificate of occupancy.

If you receive a complaint, you will need to go to an administrative hearing or a court. This procedure will be described in the Chapter 11, Going to Court.

AVOIDING TROUBLE

- Don't attempt to do a job you aren't qualified to do.
- Fix what is wrong immediately after being told to do so by the building official.
- Appeal the building official's decision if you don't agree with it and you have a legal reason for your position.
- Call the building official if you get a notice of violation so you can fix what's wrong by the deadline imposed.
- If you need more time than the building official gives you, ask for an extension of time, but be reasonable.

CONCLUSION

If you respond promptly to a notice of violation, you may be able to avoid having to go to a court or an administrative hearing. You can do this by complying with the order of the building official or by appealing the decision. If you need more time, don't hesitate to ask for it, but make sure your request is reasonable.

CHAPTER 10

APPEALING THE DECISION OF A BUILDING OFFICIAL

INTRODUCTION

If you do not agree with the decision of the building official, you may appeal her decision. Until your appeal is decided, you will not have to comply with the order of the building official, but you must not disregard it either. The board of appeals will decide how you must respond to the order. But, you must follow the proper procedure in filing your appeal so it will not be denied on a technicality.

BASIS FOR THE APPEAL

To have any chance at being successful, you must have a reason for the appeal. These are considered acceptable grounds under the International Building, Fire, and Residential Codes:

- The true intent of the code or the rules legally adopted under the building code has been incorrectly interpreted.
- The provisions of the building code do not fully apply.
- An equally good or better form of construction is proposed.

These grounds are very limited. You must show that the building official incorrectly interpreted the building code. You may need your own expert to contradict the building official's opinion or to show that the second basis applies, that the provisions of the building code do not fully apply. You may have more success arguing the third point, especially if

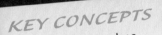

you are using a new product or method and can back up your position with research or testing.

The fact that it may be expensive to do what the code requires is not a basis to appeal the decision. Safety costs money. If you're not prepared to do what the code demands, you should rethink the project. If you don't have a proper basis for appeal, don't waste your time. Fix the problem.

In the Field

"The only appeal we ever had was a man who had converted his single-family home into three apartments without a permit and in violation of the zoning code. He tried to contend that because there was only one kitchen that everyone shared, it was still a single-family home. However, I had evidence that each unit had separate locks and mailboxes and the tenants didn't share household expenses. He tried to evict one of the tenants and she gave me a copy of her lease. The Board of Appeals was not impressed with his argument that everyone lived together as one happy family, and he had to reconvert the house back to the way it was and pay fines for doing the work without permits."
—*Stacey, Code Inspector* ■

PRIOR TO FILING THE APPEAL

Before you go through the trouble of appealing the decision of the building official, talk to him and find out if you can come to an agreement about your differences. There are always new methods of construction coming on the market, especially with the rise of "green" construction, and it may be an opportunity to educate the building official about something new. Of course, if you never should have done what you did, the conversation should be about fixing the problem.

APPEAL PROCEDURE

If you do decide to appeal the decision of the building official, you need to do it in writing and you have to follow the time requirements in the building code. For example, if you have to file an appeal within 20 days from being served with the notice of violation, you have to file it within 20 days, not on the 21st day. Once the time runs out, so does your right to appeal.

The following is an example of what a notice of appeal looks like.

Name of Appellant: _____

Address of Appellant: _____

_____

Basis for the appeal: (check all that apply)
- ☐ The true intent of the code or the rules legally adopted under the building code has been incorrectly interpreted.
- ☐ The provisions of the building code do not fully apply.
- ☐ An equally good or better form of construction is proposed.

Explanation (describe why you believe the Board of Appeals should reverse the decision of the building official):

_____

Supporting documents attached: [ ] Yes [ ] No

By _____

Date filed: _____

Hearing date set for: _____

Notice of hearing sent: _____, 20 ____

The name of the person or entity receiving the notice of violation.

Address of the person or entity receiving the notice of violation

The appeal must be based on at least one of these grounds.

Describe why you believe you have a basis for appeal.

If you have documents, you should check the box marked "yes." If you do not have documents, check the box marked "no."

Signature of the party appealing.

This should be filled in by the local jurisdiction and must be a date within the time set for appeal by the ordinance.

This should be set by the local jurisdiction.

This should be filled out by the local jurisdiction.

Figure 10–1   Example of notice of appeal.

Notice that supporting documents may be attached to an appeal. Most building officials have received considerable training in their field. If you hope to persuade the Board of Appeals to overturn the building official's decision, you must show why the building official is wrong. You may have to hire your own expert, such as a fire protection engineer, to show the Board of Appeals that you are right.

Figure 10–2 If you hope to persuade the Board of Appeals to overturn the building official's decision, you must show why the building official is wrong.

CHECKLIST—Appeals

☐ Is there a Board of Appeals you can appeal to in the local jurisdiction?

☐ Are there rules of procedure adopted by the Board of Appeals to conduct its business?

☐ If so, have you read the rules of procedure?

☐ Have you filed a notice of appeal within the time set by the statute or ordinance?

☐ If the notice is late, has the appellant been notified that his/her/its appeal is denied because of failure to meet the deadline?

☐ If the notice is timely, has a date been set for a hearing that conforms to the time limit set by the ordinance?

☐ Does your request conform to the grounds for appeal in the code? For example:

    ☐ The true intent of the code or the rules legally adopted under the residential code have been incorrectly interpreted.

continues

JUST ✔ CHECKING

CHECKLIST—Appeals (continued)

- ☐ The provisions of the residential code do not fully apply.
- ☐ An equally good or better form of construction is proposed.
- ☐ Is there a time limit set for rendering a decision?
- ☐ Have your received notice of the decision of the Board of Appeals?

This chart explains the process involved in appealing the decision of the building official.

PARTY RECEIVES NOTICE OF VIOLATION FROM BUILDING OFFICIAL
↓
PARTY FILES NOTICE OF APPEAL WITH LOCAL JURISDICTION'S BOARD OF APPEALS
↓
HEARING DATE SET
↓
PARTY MAKES PRESENTATION BEFORE BOARD OF APPEALS
↓
BOARD OF APPEALS MAKES DECISION
↓   ↓
BOARD UPHOLDS ORDER,   BOARD REVERSES ORDER
                             MATTER CLOSED
↓
PARTY MUST COMPLY,
APPEAL TO CIRCUIT COURT OR
BE SUBJECT TO NOTICE OF VIOLATION

Figure 10–3 Appealing the decision of a building official.

Unless your appeal is based on a very simple matter, it is wise to hire an attorney for the appeal.

The hearing is usually before a Board of Appeals that is made up of a number of individuals from the community who may have special qualifications. The rules of evidence are not as strict as they are in a courtroom. You will be given a chance to present your evidence and ask questions of any witnesses. While appealing to the emotions of the board members to make them feel sorry for you may seem like a good idea, it's not a way to win your case. The board members must uphold the standards set forth in the code. They can't deviate from them because you are a likable person with a hard-luck story. Don't try to save money if you are going to appeal when you have a lot at stake in the outcome. Spend the money on a good attorney and follow her advice.

Decision Of The Board Of Appeals

The decision of the Board of Appeals will usually be sent to you in writing after the members have had a chance to deliberate. If they agree with you, the notice of violation no longer applies to you, but, if they do not, you must either comply or appeal the board's decision to the local court for an administrative review. This is a costly and time-consuming process, and you have to decide if it's worth it. You also have a very strict time limit under state statute to follow. You'll definitely need an attorney if you choose to appeal the decision of the Board of Appeals.

AVOIDING TROUBLE

- Call the building official and see if you can work out an agreement about what he wants you to do.
- Consult with an attorney and see if you need her help.
- File the notice of appeal within the time limits set by the statute or ordinance.
- Make sure your reason for appealing is based on standards that are in the statute or ordinance.
- Be prepared for the hearing by having all of your witnesses and evidence present.
- If you lose, comply with the building official's decision or appeal the case to the local court within the time set by state statute.

CONCLUSION

If you believe you have a legal basis for an appeal of the building official's order, you may be able to appeal it to the local Board of Appeals if the local ordinance provides for that procedure. If you decide to do so, carefully follow the rules for appeal. Make sure you are prepared for the hearing by bringing with you the necessary witnesses and/or evidence. If the issue is complicated, you may want to hire an attorney. If you lose the appeal, be prepared to comply promptly with the building official's order.

CHAPTER 11

GOING TO COURT

INTRODUCTION

If you've never had to appear in court before, it can be a nerve-wracking experience. You are in a room full of strangers worried about the outcome of the ticket brought against you. Some people will choose to hire an attorney just so they don't have to go through this process alone. This chapter will describe the procedure that you can expect to encounter during your court visit.

VENUE FOR THE HEARING

When you have a building code violation, there are different types of hearings that determine where it will be heard. The first type of hearing is called an administrative hearing and the second is a trial in a courtroom.

Administrative Hearings

Certain state laws allow local jurisdictions to set up their own hearing processes. An administrative hearing usually takes place at the city hall of the local jurisdiction bringing the charge. The person hearing the case is not a judge but a hearing officer, usually an attorney. This person normally has the ability to levy a fine against you and demand that you correct the problem if you are not yet in compliance. The rules of evidence are more relaxed and the hearing is more informal than what takes place in court. There will usually be a police officer there to act as a

KEY CONCEPTS

trial—a judicial examination, in accordance with law of the land, of a cause, either civil or criminal, of the issues between the parties, whether of law or fact, before a court that has jurisdiction over it.

Black's Law Dictionary

bailiff. The inspector often appears without a prosecutor to present the case. The inspector presents his evidence first, and then you are given a chance to introduce your evidence before the hearing officer.

At the hearing you will be allowed to call witnesses on your behalf and to present evidence such as pictures or documents to the hearing officer. Once the hearing officer has heard both sides, she will usually then render a decision, giving you time to pay the fine or comply. It is possible for the hearing officer to dismiss the ticket depending on the nature of the evidence.

If you lose the case before the hearing officer, your only remedy is to appeal the decision to the local circuit court, usually in an administrative review action. You will definitely need an attorney to help you with such an appeal. Often the cost of filing the lawsuit is more than the fine levied against you, so few people appeal the decision of the hearing officer.

Trials

The other place where building code cases are heard is in a courtroom. In a courtroom setting the person hearing the evidence is a judge or a jury. The judge has the power to issue a fine against you and order you to comply with the code. Unlike an administrative hearing officer, the judge has the additional power of contempt of court. This means that if you refuse to obey the judge's order, the judge can fine you or put you in jail until you comply with the code.

When you go into a courtroom, the court personnel present will include the judge, a court clerk, a bailiff or deputy sheriff, a prosecutor, and the witnesses against you. Just as at an administrative hearing, the prosecutor presents the case on behalf of the local jurisdiction by calling the inspector as a witness. You will have an opportunity to present your evidence before the judge makes a decision.

COURT PROCEDURE

Arraignment

When you go to a courtroom the court clerk calls the cases by the name of the party or case number. When your name is called, you then go before the judge and the court will ask if you plead guilty or not guilty. This is what is called an arraignment. You may have seen this on television. For example, whenever a celebrity gets arrested there is a big fuss made over the first court date, the arraignment, but

KEY CONCEPTS

arraignment—the formal act of calling the defendant into open court, informing him of the offense with which he is charged, and asking him whether he is guilty or not guilty.

Illinois Compiled Statutes

it is really a very simple procedure. Once you have entered your plea, the court will then either sentence you if you plead guilty, pass the case for a trial if you've pleaded not guilty, or continue the case for trial depending on the way the judge is controlling her docket.

Plea Bargaining

Often when the case is passed for trial, the court may give you an opportunity to have a plea bargaining conference with the prosecutor and the inspector. Plea bargaining has a bad reputation, but it is really the only way that the court system can accommodate the number of cases brought before it every day. The positive thing about plea bargaining, from the standpoint of the defendant, is that you know exactly what your sentence is going to be and you don't have to worry about whether the judge is going to do something worse to you. From the prosecutor's standpoint, plea bargaining is a positive thing because it gets cases off of the trial call and it enables the prosecutor to get an agreement for compliance, which is more important than any other outcome, even a large fine.

If there is some hardship regarding your case, this is the time to bring this to the attention of the prosecutor. For example, if there are problems on your property but the house is in foreclosure, this is information that the prosecutor should know. It may be impossible for you to comply depending on your financial situation, and the prosecutor may agree to wait for enforcement until the bank is in possession of the residence or until you can secure financing to correct the problem. Sometimes the building official knows about local resources that may help you, depending on the nature of the problem.

The plea bargaining conference is not the time to complain about the inspector or the local jurisdiction unless you have some very specific facts that would get you sympathy from the prosecutor because of some misunderstanding that occurred that was not your fault. Usually the prosecutor will have a very jaded attitude about defendants because by the time you get to court, it means that you have ignored the notice of violation from the inspector and have not been interested in resolving the issue short of a hearing.

During the plea bargaining session, you and the prosecutor may discuss the facts of the case in order to determine whether there is any advantage to either side proceeding to trial. Building code cases usually do not end up in a trial because the evidence is clear-cut—either you worked without a permit or you didn't. Either you had a certificate of occupancy or not. Therefore, the only thing you really

KEY CONCEPTS

plea bargaining—a conference between the prosecutor and the defendant or the defendant's attorney during which the parties try to reach an agreement as to the ultimate disposition of the case.

need to discuss is how to resolve the matter to the satisfaction of both parties. The prosecutor may offer to recommend a lower fine if you plead guilty and promise to comply with the code by a certain date. You should pay attention in the courtroom to hear what kind of fines the judge is giving to other people to get an idea as to whether the fine offered to you is within a similar range. The plea bargaining conference is also an opportunity to clear up any misunderstandings. For example, if you can show the prosecutor that you don't own the property and are not legally responsible for the ticket brought against you, you may be able to get your case dismissed.

If you are able to come into compliance with the code prior to the court hearing date, you should bring this to the attention of the prosecutor. In some jurisdictions, the prosecutor dismisses the case upon compliance. In others, it is the basis for a reduction in the amount of the fine levied against you. If you can't come into compliance by the date you are in court, come in with a plan for compliance that you can show to the prosecutor. If you can demonstrate that you have a contractor ready to complete the work or that you have applied for the proper permit or called for a final inspection, you may be able to get a continuance for a period of time in order to fully comply. Prosecutors like defendants to be cooperative and not obstinate when there are no facts in dispute. If you don't have a legal leg to stand on, don't be difficult in a plea bargaining conference. Try to get the least penalty possible by being agreeable.

Be truthful during the plea bargaining session. It is normal to make excuses when you have to go to court. Often defendants come in during a plea bargaining conference and start to complain about how they would gladly have complied with the code if only the inspector had called them and let them know there was a problem. Good inspectors always send a notice of violation before bringing someone to court. Their files will have the green certified mail receipt, which shows that not only did the defendant receive a notice of violation prior to court, but the person actually signed for the mail. This is very embarrassing for the defendant because it looks like he is trying to be deceitful. This is not helpful for the defendant's case. Some judges will impose a higher fine when people lie in court. Prosecutors take this into consideration when making plea offers.

Going To Trial

If you are unable to reach a plea bargain, you should make sure you either have sufficient evidence on your side or a lawyer. Prosecutors deal with cases all the time and know the value of a case. They can look at the evidence and decide whether it's worth pursuing to trial or not. If after a plea bargaining conference the prosecutor is willing to go forward against you at trial, that is not good news for you. It usually means that she has sufficient evidence to obtain a finding of guilty and that you haven't raised a legal defense. For example, a legal defense to a charge of not obtaining a permit is not "I don't have the money," "I didn't know I needed a permit," or "It's not fair." Most ordinance violations are what are considered strict liability offenses. That means that it doesn't matter that you didn't intend to violate the law, you didn't know what the law was, or some other type of excuse. The only issues are whether you are responsible for the violation that is described in the ticket and whether there is a violation of the code. If the prosecutor can prove these elements, you are going to lose if you go to trial.

The judge will usually give you a continuance in order to hire an attorney or to obtain evidence if you didn't bring it with you into court that day. Ask for a new date for that purpose at the arraignment.

The accompanying flowchart describes the procedure that is followed prior to the trial taking place.

Figure 11–1 Court procedure—prior to trial.

Evidence At The Hearing

Evidence is anything that the court considers at the hearing. It can be your testimony as to what you observed; it can be photographs that either side presents into court. Evidence can be documents such as letters or copies of deeds. A photograph is admissible as long as it truly and accurately depicts the subject matter therein. If you take a photograph, make sure you take a photograph that actually shows the area in question. Too often defendants tend to take a photograph that puts them in the best light. You can anticipate that an inspector will have taken a picture from the location that actually shows the nature of the violation in your case. If that place is not in your photograph, the judge may think you are trying to hide something and deceive the court. This is not a position that you want to be in.

You can testify to anything that you observed or told the inspector, and you can testify to anything that the inspector told you if the inspector is in court. However, if you obtained information from a nameless person on the telephone when you called city hall, the prosecutor will probably object on the basis that this is hearsay. Hearsay is trying to introduce a statement for the truth of the matter asserted. If the person is not in court, he cannot be cross-examined by the prosecutor. Any statement made is hearsay and not admissible. This means that you can't get an affidavit from your neighbor and introduce it at the hearing if your neighbor can't come to court.

Photographs and videos are used all the time in court, but you need to make sure that the court has the ability to play a videotape, CD, or DVD before the court session. You can usually find this out by calling the main courthouse and asking for the chief judge's office to find out whether those types of electronic equipment are available in the courtroom that you will be in.

Figure 11–2   Judge ruling.

It is not a bad idea to stop by the court and watch the court proceedings prior to the date that you are scheduled for court. Any citizen is entitled to sit in any courtroom in this country for as long as he or she would like to. You have the constitutional right to watch other people's cases. This way you will be familiar with the procedure that the judge follows and gain a little insight as to how the judge may respond to your case.

If your defense is that somebody else is responsible for the violation, then you should bring evidence of that to court. If you have copies of deeds, contracts, and letters that you received from

Figure 11–3   You have a right to watch other people's cases. This will help you to become familiar with the procedure and how the judge may respond to your case.

the local jurisdiction, bring those too. You must have everything organized because judges do not like to sit and wait while you are looking for a particular photograph or document. Organize your evidence before going to hearing so you can present it without delaying the proceeding.

RETAINING AN ATTORNEY

You should have a lawyer if you face the possibility of severe penalties, such as large fines or if there is a possibility you could go to jail. If the case is complicated and you don't understand why the inspector has brought you to court, you probably need a lawyer. If you are nervous about presenting your own case, you should hire a lawyer to speak for you. Normally lawyers charge a set fee or by the hour in these cases. You should have a clear understanding as to how much it is going to cost you for a court appearance and any other type of work that the attorney does for you before hiring her. You should also ask the attorney what her experience is in handling these kinds of cases, because very few attorneys handle ordinance violations, especially building code matters. You may find that an attorney who practices criminal law would be better able to represent you than someone who does civil litigation in these particular types of cases.

Figure 11–4 Ask your attorney what his or her experience is in handling ordinance violations, especially building code matters. Few attorneys have significant experience in these types of cases.

TRIAL PROCEDURE

During a trial the burden of proof is on the local jurisdiction. Therefore, the prosecutor will call her witnesses first. The witness must take an oath to tell the truth prior to testifying. Usually the witness called is the inspector who handled your case. The inspector will be led through a series of questions to describe what she saw on your property and when it happened. There may be photographs introduced or other documents that help the prosecutor prove the local jurisdiction's case against you. When the witness has finished testifying, you will have an opportunity to cross-examine that person. You may ask any questions that are relevant and material to the charge against you. This is not the time to argue or to make your case. It is only the time to ask questions.

After you have finished questioning the witness, the prosecutor may conduct a redirect examination of the witness to clarify any issues raised by your questions. After the prosecutor has introduced all the evidence on behalf of the local jurisdiction, it will then be your turn. If the local jurisdiction has not made a sufficient case despite its

FIGURE 11–5
Usually the building inspector who handled your case will be called in as a witness.

evidence, you could move for a motion for a directed finding. This is a motion that asks the judge to dismiss the case because the prosecutor has not introduced enough evidence to sustain the complaint that was brought against you. It is rarely granted and so most people do not even ask for it, especially if the prosecutor has made what is called a "prima facie" case. This means that the prosecutor has produced evidence on every element of the charge against you.

When it is your turn you may testify or question your own witnesses. Or if you have hired an attorney, she will conduct direct examination of the defense witnesses. After a witness for the defense has testified, that person is subject to cross-examination by the prosecutor. It is possible that the prosecutor could ask questions of the witness to prove the case against you. That is why it is important to know exactly what the witness has to offer by way of testimony. After the prosecutor has finished cross-examining your witness, you then have an opportunity to do a redirect examination, that is, asking questions to clarify matters brought up during cross-examination. After you have completed your case, the prosecutor gets one more chance to call rebuttal witnesses against you to make her case. You will again have a chance to cross-examine any rebuttal witness, and then the prosecutor may conduct a redirect examination of that witness.

Either side may object to the evidence of the other party by saying "objection." After that, the judge will either allow the evidence by overruling the objection or deny it by sustaining the objection.

> *KEY CONCEPTS*
>
> **examination**—the series of questions put to a witness by a party to the action or his counsel, for the purpose of bringing before the court and jury in legal form the knowledge which the witness has of the facts and matters in dispute, or of probing and sifting the evidence previously given.
>
> *Black's Law Dictionary*

The rules of evidence are very complicated, and only a lawyer will know when to object and what the proper basis is for objections.

At the end of the evidentiary case, each side gets a chance to make a closing argument in the matter. The prosecutor goes first and tells the judge why she believes she has brought sufficient evidence against you. This is your chance to explain to the judge why you believe you should win the case and have it thrown out. The prosecutor gets one more chance after you finish to argue in rebuttal. After this, the judge makes her ruling.

The accompanying chart outlines the basic procedure followed during a bench trial.

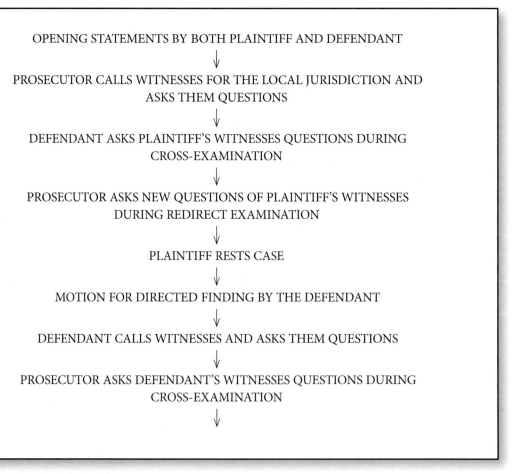

OPENING STATEMENTS BY BOTH PLAINTIFF AND DEFENDANT
↓
PROSECUTOR CALLS WITNESSES FOR THE LOCAL JURISDICTION AND ASKS THEM QUESTIONS
↓
DEFENDANT ASKS PLAINTIFF'S WITNESSES QUESTIONS DURING CROSS-EXAMINATION
↓
PROSECUTOR ASKS NEW QUESTIONS OF PLAINTIFF'S WITNESSES DURING REDIRECT EXAMINATION
↓
PLAINTIFF RESTS CASE
↓
MOTION FOR DIRECTED FINDING BY THE DEFENDANT
↓
DEFENDANT CALLS WITNESSES AND ASKS THEM QUESTIONS
↓
PROSECUTOR ASKS DEFENDANT'S WITNESSES QUESTIONS DURING CROSS-EXAMINATION
↓

Figure 11–6 Bench trial procedure.

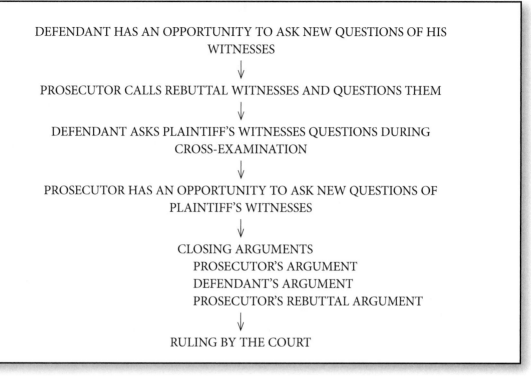

DEFENDANT HAS AN OPPORTUNITY TO ASK NEW QUESTIONS OF HIS WITNESSES

↓

PROSECUTOR CALLS REBUTTAL WITNESSES AND QUESTIONS THEM

↓

DEFENDANT ASKS PLAINTIFF'S WITNESSES QUESTIONS DURING CROSS-EXAMINATION

↓

PROSECUTOR HAS AN OPPORTUNITY TO ASK NEW QUESTIONS OF PLAINTIFF'S WITNESSES

↓

CLOSING ARGUMENTS
PROSECUTOR'S ARGUMENT
DEFENDANT'S ARGUMENT
PROSECUTOR'S REBUTTAL ARGUMENT

↓

RULING BY THE COURT

Figure 11–6   *continued*

JURY TRIALS

A jury trial is a trial in which 6 or 12 ordinary citizens hear the evidence against you and at the end of the trial deliberate in the jury room to decide whether you are guilty or not. Most people do not demand jury trials in these kinds of cases because they lack the skill to conduct a jury trial. There are special instructions that must be drafted in any jury trial, and judges do not like courtroom time being taken up with jury demands when the case could have been resolved at a bench trial before the judge. It is possible that the judge could impose a higher fine if you have a jury trial rather than a bench trial. You do have constitutional right to a trial, but the reality is that judges like to have jury trials on fairly serious matters, especially if there is no material issue in dispute, which is usually the case in building code violations. If you are going to demand a jury trial in your case, you should hire a lawyer as soon as possible.

AFTER THE HEARING

After all the evidence has been introduced and the judge has heard closing arguments, she will make a decision as to whether you are guilty or not guilty. If you are found not guilty, that is the end of the case and you are free to go at that time. If you are found guilty, the judge may sentence you on that day or at a future time depending on the local court procedure. You should find out ahead of time what the possible penalties are in your situation by consulting with a lawyer if necessary. It is possible that you might receive a fine and be ordered to comply with the code if you are still not compliant. If you are given a fine, the judge will usually give you time to pay the fine. If the judge orders you to come into compliance, there will usually be a future court date that you will have to attend in order to prove to the judge that you have obeyed her order. Sometimes the court may defer sentencing until you come into compliance and then give you a lower fine. This book cannot tell you how much your fine will be because that varies from jurisdiction to jurisdiction. Usually though, the judge can impose a separate fine for every day that the violation occurs. It is not unusual for people to run up fines in the thousands of dollars if the problem persists over a long period of time.

You may be given a day to come back to court to prove that you complied with the judge's order. If you fail to appear, some judges will issue a warrant for your arrest, so make sure you know what your next court date is before you leave the courtroom. If you have to come back to show you have complied with the judge's order, be truthful.

In the Field

"Digital cameras have really changed the way we do business. I always stop by the job site on the morning of the court hearing and take pictures. That way, I can report to the judge whether the defendant is in compliance with a court order or not. I can't believe how many people lie to the judge and tell him that they are in compliance. I always let my prosecutor know ahead of time whether a defendant complied with the judge's order. It causes quite a commotion when the prosecutor asks for the photographs I took that morning and shows them to the judge after the defendant has said there's been compliance. The defendants just stand there in shock." —*Joan, Building Inspector* ■

AVOIDING TROUBLE

- When you go to court, bring all of your evidence with you.
- Ask to have a conference with the prosecutor to try to resolve your case.
- During the plea bargaining conference, be respectful to the prosecutor and the inspector even if they are not being respectful to you.
- If you go to trial, don't interrupt or make comments.
- Come into compliance with the code prior to your court date.
- If you are in compliance, ask the prosecutor if it is possible to dismiss the case against you or lower the fine.

CONCLUSION

Going to court or to an administrative hearing can be an anxiety-provoking experience. You can reduce your concerns by being prepared for a trial or hearing. If you have the opportunity to speak with the prosecutor before trial, you may be able to reach an agreement by plea bargaining your case. Unless your case is fairly simple, it would be wise to seek the help of an attorney. When in court, make sure that you observe the proper courtroom decorum, especially during trial.

CHAPTER 12

AVOIDING BUILDING INSPECTORS' PET PEEVES

INTRODUCTION

Building officials and inspectors are human beings just like you, with the same emotions and frustrations all of us experience. Consequently, how you interact with them can make a difference in how smoothly your project progresses through the permit process. Some inspectors are very knowledgeable and hardworking. Others are not as earnest and are impressed with the authority of their position. This chapter will discuss strategies to avoid problems in dealing with building inspectors.

TRAITS ADMIRED BY BUILDING INSPECTORS

This is truly an area in which following the Golden Rule is beneficial for you. Treat the building official or inspector with the same respect you expect. In addition to treating the inspector with courtesy, there are other behaviors you should emulate.

The same standards that you admire are shared by building inspectors. Honesty is one of those values. Being respectful to the staff is another trait that building officials and their support staff remember. Building officials value competency, so following the code will earn the respect of the building inspector. Following are some of the characteristics of owners and builders that are guaranteed to get positive attention. They are based on actual interviews with building officials and inspectors.

Honesty

Develop A Reputation For Honest Dealing

Unfortunately, people lie to inspectors all of the time. No matter is too small to lie about. They lie about leaving phone messages, and they lie about phantom conversations they had with city employees. They lie about when they will submit an application and why they couldn't follow through on a promise. By being truthful in your dealings with the building official, you will set yourself apart from those whose reputation for honesty is tarnished. Anything those persons do or say will always be suspect and they will never get the benefit of the doubt, whereas you will be the kind of person that is of no concern to the building official.

Be Truthful Regarding Notices Of Violation

If you have received a notice of violation, admit it when having a conversation with the inspector or the prosecutor. Unfortunately, defendants deny they ever received notice of the violation in court all the time. There is usually a paper trail that shows when the notice was served and who received it. If an employee or family member signed for the notice and never passed it on to you, recognize that it's not the inspector's fault if your staff or a family member didn't pass on this vital information to you. Express your desire to

Figure 12–1 Don't distract the inspector during an inspection.

fix the problem. The inspector will understand business or family communication mishaps.

Allow The Inspector To Perform The Inspection Without Distraction

Inspectors report that as soon as someone tries to distract them while they are conducting an inspection, they suspect someone is trying to hide something. Inspectors usually have a very specific way of conducting an inspection so they don't miss anything. Do not try to deter them from conducting the inspection their way.

Competency

Educate Yourself About The Building Code

Before undertaking a construction project, make sure you understand what the requirements of the code are. It is unwise to challenge the expertise of the building official unless you have special knowledge and can point out to the building official why your interpretation of the code is correct. This does not mean that you shouldn't ask for a copy of the building code section in dispute if the building official is relying on a section with which you are not familiar. If the code says what the building official is telling you it does, don't argue. You may be able to access the building code on the Internet by searching for "code" or "code of ordinances" and the name of the local jurisdiction.

Hire Competent General Contractors Who Possess The Necessary Skill To Do The Job

Too many people want to save money by doing a project themselves or acting as their own general contractor. You may be able to do this if you are organized and willing to learn. However, inspectors deal constantly with people who don't have a clue as to what they are doing. Sometimes out of frustration, a building official will help the homeowner do a sketch for something like a fence that is necessary for the building permit in order to get it approved. But, you should never expect the building official to do what you can hire someone to do.

Call To See If A Permit Is Necessary

Chapter 3 describes the problem of failing to obtain the necessary information regarding permit requirements. People can save themselves so much grief if they call the building department to see if they need a permit for their project. When they don't call, it ends up costing them more time, aggravation, and money than if they'd just taken the time to contact the building department.

Hire Design Professionals Who Know The Building Code

You expect that when you hire a professional, she will know what she is doing. Unfortunately this is not always the case. When plans are rejected, the design professional may blame the building official in order to cover up her ignorance about the building code. This is why it's important to get a competent referral when hiring a professional. The building official can't tell you whom to hire and who is competent or not, but you can ask the building official if he knows a particular person. Watch his reaction to see if you can read between the lines. Check the references of the design professional before engaging his services.

Figure 12–2 Hire a competent architect.

Obtain Permits

Obtaining the right permits for your project is the single most important thing that will ensure a positive relationship with the building department. Failure to obtain a permit violates the code and turns inspectors into law enforcement officers, which they would rather not be. Projects without permits will usually have numerous building code violations. This takes up a lot of the inspector's time that she could be using to perform inspections for people who followed the law.

Submit Complete Applications

People often complain about how long it takes to get an application approved. However, a huge factor is the applicant's failure to submit the necessary construction documents or a complete application. This failure slows down the plan-review process. If the building department has to return the application with comments and it has to be resubmitted, it's going to take longer to get a permit. Therefore, have all of your information and documents in order before submission.

Figure 12–3 Fill out the application completely.

Don't Begin Work Before The Permit Is Issued

Good builders begin work after the permit has been issued, picked up, and placed on the job site. Until the permit is in hand and properly posted, no work should begin. Don't rush the project. Give yourself a reasonable timeline so that you don't feel pressured to begin before you have the permit.

Leave Work Uncovered Before It Is Inspected

The inspector needs to see the work done before he can approve it. Keep the walls open until the necessary inspections are performed. Covering up the work can be interpreted as an attempt to hide something. A person can be forced to open up the wall so the building inspector can complete the inspection. Any cost involved is the responsibility of the builder, not the local jurisdiction.

Call For The Necessary Inspections In A Timely Fashion

Call when it is time for an inspection. Building inspectors are too busy to monitor a permit holder's building project. They rely on you to call for an inspection at the proper time. You'll be able to get a certificate of occupancy by having the proper inspections, so follow the code and schedule your required inspections.

Competent contractors call in a timely fashion for an inspection and don't wait until 3 p.m. on a Friday and expect the building inspector to drop everything to attend to the needs of the permit holder. Building inspectors have very full days with many inspections. If the builder calls too late, he's going to have to get in line and wait his turn.

In the Field

"It doesn't happen often, but once in a while, a particular contractor calls in the afternoon to schedule an inspection for a footing and says something to the effect that, 'If your inspector wants to see it before it is poured, let him know it will be ready at 10:00 a.m. tomorrow.' This arrogant attitude shows a lack of 'team playing' and cooperation, and a disregard for the fact that we cover over 700 square miles and may well be booked-up at jobs 45 minutes away, disregard for the code requirement to have it inspected, and disregard for the requirement that he give us notice. My response might be, 'I can get the inspector there by 1:00 p.m.' (which is only by juggling things around and pushing the inspector and hoping things go well). Then, I find out later that he poured the concrete before the inspector arrived at 12:15 p.m.! So, on top of the insults and the attitude and then our efforts to accommodate him, in spite of all else, he wastes our time!" —*Eric, Building Official* ■

Build Only What You Can Afford So You Meet The Requirements Of The Code

Do research to make sure that the cost of the project is feasible. If a person can't afford to meet the building code, the project shouldn't be undertaken. High cost is no reason not to comply with safety standards. The building and fire inspectors will usually work with parties to see if there are other alternative methods that achieve the proper result at a lower cost. But, it is the responsibility of the owner to do the research before moving forward with the project.

Do Not Change The Plans Without Consent

The staff that reviews plans takes time and effort to make sure everything was in order and complied with the code. The permit holder has a permit for only the work approved. Changes are improper and can result in a violation notice. If you need to change something, make sure you get approval in writing. You don't want to get caught doing something without the permission of the building official.

Obtain A Certificate Of Occupancy Before Moving Into A Building

Moving in possessions or people can create tremendous hardship if caught. The occupant may have to remove the contents of the building because of not waiting for the certificate of occupancy. Until there has been an approved final inspection, no certificate of occupancy can be issued. If you are buying a property with new construction on it, demand to see the certificate of occupancy before completing the sale.

Fix Problems Quickly

When an inspector finds a problem, fix it quickly. People make promises to building inspectors all of the time, agreeing to correct deficiencies, but do not follow through on their assurances. It is not the job of the building inspector to keep reminding the permit holder to follow through with promises. If the problem can be fixed by most people within a day, it shouldn't take two weeks to resolve the matter.

Fix Inoperable Fire Alarms Quickly

Hire an alarm monitoring company to make sure fire protection systems are operating properly. However, you can't blame the failure to keep the fire alarm in good working order on the monitoring company. If it fails to fulfill its obligations, you are responsible because you are the one that hired them. Fire alarms that are out of service put people's lives and property at risk. Fixing a malfunctioning fire alarm should be given top priority.

Figure 12–4   Fix inoperable fire alarms quickly.

Respond To A Notice Of Violation Immediately

The purpose of a notice of violation is to get the attention of the responsible party so the problem can be corrected. The building official appreciates diligence and will work with the permit holder to resolve the issue. The violation will not go away if ignored. Always call the building official to find out what must be done to come into compliance.

Obey Stop Work Orders

Most stop work orders are given only because there is a true problem at the job site or because the project lacks the necessary permits. This is a serious matter that can get the permit holder and workers into big trouble with hefty fines. If you are served with a stop work order, obey its conditions. Find out what you need to do to get it lifted and then do it.

Courtesy

Be Respectful Of The Inspector's Schedule

Be respectful of the inspector's time. Unfortunately, too many permit holders berate the inspector when he can't come when the permit holder wants him to do the inspection. Inspectors usually aren't sitting by the telephone waiting for permit holders to call for an inspection. They are out in the field trying to get as many inspections done as possible. If the inspector can't come when you want him to, call sooner next time.

Don't Complain About The Cost Of Complying With The Code

Only undertake a project if it's feasible to follow the code and keep costs under control. A person who chooses to get involved in the project must do the necessary research to get a reasonable estimate of what the project is going to cost following the local building code. It's not the inspector's fault if the cost of following the code makes the project impossible because of the expense.

Treat Staff With Respect

Building inspectors do speak with their support staff. Every person in the department will talk about what a jerk a person was if that person argued with someone or lost his temper. That person's application probably won't receive top priority either. By forming a respectful relationship with the support staff, you may even get special treatment.

Accept Responsibility For Mistakes

If you make a mistake on the project, take responsibility. There are few things that can't be fixed. If you blame the building inspector for not getting a certificate of occupancy because you waited too long to get the final inspection, you're missing a valuable lesson in time management.

Take Steps To Avoid Damaging Neighboring Property

Building officials don't like to be referees. They hate having neighbors complain to them about the building project next door because it means added headaches for them. When they get a citizen complaint, they have to take action. They don't want the neighbor to complain to their boss. If your construction creates a problem on the neighboring property, resolve the issue quickly by telling the owner how you're going to fix what you did and then do it. People understand that problems happen, but how problems are resolved is critical.

Follow Though With Promises

There will always be problems during the construction process. Building inspectors understand that and just want the permit holder to follow through with any corrections that must be done. If you promise to call the building inspector or drop plans off at a certain time, do it, or at least call the inspector so she knows what's going on. Good communication is the key to a successful relationship with the building inspector.

Don't Ask For Charges To Be Dropped

Asking for charges to be dropped shows disrespect for the time it took the building official to write the notice of violation and the time spent preparing for court. Why should the building official give you a break if you ignored the notice of violation? An exception to this rule is if you know that the regular practice of the local jurisdiction is to dismiss cases if there is compliance.

Don't Say: "No Other Town Requires Us To Do This"

Remember when you told your mother that none of your friends had to follow such stupid rules and she told you that as long as you lived in her house you'd follow them? The same principle applies to building codes. Just because other towns aren't strictly enforcing the code doesn't mean that the town you are in shouldn't.

Don't Say: "You've Never Required This Before"

This often happens when a new building official is hired and seeks to bring more professionalism to the office. The days of a wink and a nod are over and the code is being strictly enforced. Perhaps a builder was fortunate in the past to escape scrutiny. But with a new "sheriff" in town, everyone must comply with the code.

In the Field

"I once had a builder tell me that he always made sure that he had his best people supervising jobs in the town I was prosecuting for because it had a tough enforcement reputation. He said knew he could slide in the bordering towns. I'm sure the building inspectors in those towns wouldn't be happy to hear that."
—*Christine, Village Prosecutor* ■

Ethical Choices

Don't Call Politicians To Intervene In Your Cases

Unless the mayor is a permit holder's last resort, it is unwise to bring politics into the picture. It is tempting when a person knows people with influence in local government to call them for help, but it doesn't promote public safety and will often backfire. The best course of conduct is to follow the chain of command by beginning with the inspector, his supervisor, and so on. The person who uses political influence may get his way on a job by using his connections, but what happens in the future if the political landscape changes?

Never Offer Bribes Or Gratuities

It's a crime. The person offering the bribe could go to jail. Is a project worth ruining one's life? Never put a building inspector in the position of having to call law enforcement.

In the Field

"This builder needed to vent some fans in the building. He got out of his car and said, 'How much is this going to cost me?' I said, 'What are you talking about?' He told me that in the city nearby the rate was $25 a fan. I told him he wasn't in that city, and I didn't want to hear anymore of that kind of talk." —*Tim, Building Inspector* ■

Never Drop Names Of Important People

This will definitely offend the building official and fail to impress him. Additionally, it could backfire.

In the Field

"I once had a lawyer who used my name when he was given a notice of violation for working without a permit and tried to get out of the situation. Of course, the inspector gave him a ticket anyway. I wasn't even friends with this guy. I told him dropping my name was a bad move in that I had instructed all of the law enforcement officers in the local jurisdiction to give a ticket to anyone who dropped my name in a conversation when they were trying to talk themselves out of a ticket (especially my teenage daughters). Of course, the lawyer didn't know about that!" —*Jo, City Prosecutor* ■

AVOIDING TROUBLE

- Treat every person you deal with in the building process with courtesy and respect.
- Meet all deadlines set by the local jurisdiction.
- When you receive a call or notice from the local jurisdiction, respond promptly.
- Don't expect special treatment.
- Give yourself plenty of time when arranging for inspections and when trying to obtain a certificate of occupancy.
- Follow the codes of the local jurisdiction.
- Don't undertake a project if you don't possess the necessary skill. Hire someone who does.

CONCLUSION

Following universal rules regarding courtesy, competency, and ethical behavior will guarantee that extraneous issues do not sidetrack a project. By being a person of integrity, following proper procedure, treating everyone with respect, and never engaging in unethical or criminal behavior, you will never be the subject of an inspector's pet peeve.

CHAPTER 13

BRIBERY AND CORRUPTION

INTRODUCTION

The majority of employees working for building departments are honest, knowledgeable, and hardworking. However, as in every occupation, there are dishonest people who try to use their positions for their own financial benefit or just show up to work to obtain a paycheck. Given the volume of permits issued, the bureaucracy in some local governments, the pressure to begin a project or complete it, and human nature, it is not surprising that inspectors will exploit a situation by accepting "gratuities" to expedite the process and that builders will pay bribes and consider it a cost of doing business.

Employees who are corrupt or incompetent can make life very difficult for you. Corruption is a criminal matter, but it affects only those situations in which the "fix" is. Incompetency affects everything the person does. Sometimes it's difficult to tell which is worse. This chapter will discuss strategies for dealing with both obstacles.

Figure 13–1 Bribery and corruption exists in every profession.

WHAT TO DO IF SOLICITED FOR A BRIBE

If a person is solicited for a bribe, the first thing to know is that if the person responds by agreeing to it, she is committing a crime too. Therefore, no one should participate in such a scheme unless she is doing so in cooperation with a law enforcement agency. If you give an inspector a bribe, you are just as guilty as he is for soliciting it.

Identifying the right agency to report the crime to is important. A person's first instinct may be to report the crime to the police department of the local jurisdiction. This would usually be a good place to start unless there is a climate of corruption in the local jurisdiction, where certain people can act with impunity because of their political connections.

In the Field

"Everybody used to be treated equally by our department until the new mayor took over. After that we were told to leave certain people alone even though we knew they were violating the local ordinances. I got fed up so I got a job in another city and quit. I couldn't stand not being able to do my job." —*Luke, Building Inspector* ■

If you find yourself in this type of situation and fear that going to the local authorities is fruitless or could backfire on you, you may wish to contact another agency that has jurisdiction over public corruption. These might include:

- The county's district attorney
- The state's attorney general
- The Federal Bureau of Investigation (FBI)

These law enforcement agencies may ask you to allow them to use an eavesdropping device in order to catch the offender speaking with you about the crime or accepting the bribe money. Only you can decide how far you want your cooperation to go. Most law enforcement agencies will want to catch the perpetrator in the act so there is no misunderstanding about her intentions. If the only evidence the authorities have is your accusation, it is unlikely they will charge the person unless she confesses to the crime. A more likely scenario is for the accused to deny having such a conversation, to insist you misunderstood the conversation, to say that you were the one suggesting a payoff, or that you made up the allegation because you didn't like her enforcement of the code.

Sometimes corruption is subtle. Some people seem to sail through the application process, and for others, their approval is delayed. There may be an innocent explanation, however. Builders who have gained the trust of the building department because they know how to do things right may get the benefit of the doubt because of their reputation. Other contractors who have given the inspectors problems in the past may face greater scrutiny, which in turn slows down the process.

There are times, though, when friends or supporters of local politicians get special expedited treatment because of their friendship with those in power, because of their financial contributions to them, or both. Unless a person takes steps to become an insider or to change the local political landscape, there is little one can do about this type of behavior unless the local jurisdiction has a prosecutor aggressive enough to go after such activity. Many people don't even view it as criminal, just a form of influence.

DEALING WITH INCOMPETENT INSPECTORS

While most inspectors are properly trained and knowledgeable, there are always individuals who are incompetent or just collecting a paycheck until retirement. Many states and local jurisdictions have strict standards as to who can become an inspector. In other places, there are no legal standards and the mayor's shiftless nephew could become an electrical inspector without any qualifications.

If you have difficulty with an incompetent inspector, do not lose your temper with that individual. Remember, you can always appeal her bad decision to the Board of Appeals. But, before taking that kind of action, arrange an appointment with the inspector's supervisor and hope that that person is competent and will give you the relief you seek.

Another technique to use, if the inspector is telling you to do something "because I said so," is to ask the inspector to give you a copy of the section of the code upon which she is relying. At least this way you will be able to verify that the inspector is following the code. Sometimes, though, in searching for the code section, the inspector will find that no such requirement exists or that the actual section does not say what the inspector told you it did. You should try to find the code section ahead of time by searching the Internet.

You may have to walk a tightrope in how you handle this type of situation. If you gloat that you were right, you risk retaliation by the inspector on other matters she controls, such as how quickly she shows up for other inspections. Be gracious in victory. You may be able to use this to your benefit if the inspector is less likely to challenge your interpretation the next time.

The danger in having incompetent inspectors is self-evident. They often allow projects to pass inspection that shouldn't. Public safety is in jeopardy when competent inspectors aren't hired by local government. Unfortunately, in large cities, it is not uncommon. Prize-winning journalist Carol Marin broke a story in the *Chicago Sun-Times* about the 19-year-old son of a carpenters' union leader who was given the job of city building inspector at an annual salary of $50,000, despite the fact that the city requires building inspectors to complete an apprenticeship program, obtain journeyman status, and have two years of work experience. Because apprenticeship programs typically take four years to complete, the inspector would have had to start the program at age 13. After the story broke, the inspector resigned.

DEALING WITH ARROGANT INSPECTORS

Some building inspectors are competent but overly impressed with their own power. They may be difficult to deal with if their authority is questioned. The permit holder must be professional at all times but may need to speak with the building inspector's supervisor if the building inspector is being unreasonable. Sometimes the behavior of the inspector comes out during a plea bargaining session with the prosecutor after a notice of violation has been issued. The prosecutor may be sympathetic.

In the Field

"I once dealt with an inspector who frustrated so many applicants by constantly changing the rules for them. They could never get a straight answer from him over his interpretation of the code. He'd write violations up on the basis that they had violated the 'authority of the local jurisdiction.' After I began working with him, I was very clear that I needed a copy of the code provisions he was using so I could review the charges. If he couldn't come up with a provision, I instructed him not to write the ticket. By forcing him to search for a specific section of the code, he ended up writing better tickets and becoming a better inspector. We also got fewer complaints about how difficult he was to deal with." —*Sarah, County Prosecutor* ■

AVOIDING TROUBLE

- If solicited for a bribe, don't participate unless it's part of a police investigation.
- Research the best law enforcement agency to complain to if you decide to go to the authorities regarding an unlawful solicitation.
- Keep detailed notes on every conversation that may be important.
- Know the politics of the local jurisdiction.
- Appeal the decision of the inspector if he has misinterpreted the code.
- Be professional in dealing with the inspectors, no matter how incompetent or corrupt they may be.

CONCLUSION

Thankfully, most owners or contractors never have to deal with corruption or incompetent building inspectors because most inspectors are ethical and well-trained. But, if these situations happen, it is important not to participate in an illegal act and to report the matter to the proper law enforcement agency and follow its advice. With an incompetent inspector, the permit holder may have to deal with the situation by pressing the building inspector on the section of the code at issue or by dealing with the person's supervisor.

CHAPTER 14

CONCLUSION

A construction project can often be a frustrating and anxiety-provoking undertaking. There are many things beyond the control of the builder or home-owner. Weather-related delays are not uncommon, materials get backordered, subcontractors don't show up as promised, and everything takes longer and costs more than planned for. Therefore, the last thing anyone thinking about construction or renovation wants is to worry about headaches with the local authorities over building- and zoning-code-related issues. Luckily, contractors and owners have a great deal of control over interactions with the local jurisdiction if they do the proper research and use due diligence regarding all aspects of the project.

Understanding the scope and purpose of the building code that applies to your project is a starting point. You need to find out which particular building codes govern your project. If you believe you don't need a building permit, you have seen in this book that it is important to verify this with the local building department unless the code is very clear that no permit is necessary.

You now know that many people overlook a number of important issues that could have a major impact on the success of their project, especially those that involve zoning and stormwater management. The failure to fully investigate the impact of these considerations on your construction can lead to costly mistakes, including buying property not suitable for the intended use of the purchaser. Ignoring environmental concerns can add tremendous

cost to the project if not carefully considered prior to beginning construction.

You have also learned that owning property does not allow you to build structures without restrictions. The wise person will investigate all of these special considerations when deciding whether the project is economically feasible given the requirements of the building and zoning codes. While there may be considerable disappointment in being unable to build one's dream house or to have to scale back a commercial project, it's better to know the restrictions and limitations before spending thousands of dollars.

Figure 14–1   Can you afford to build your dream house and meet the building code?

If the construction project is possible, you can sail through the permit application process by being informed and organized. Many of the problems people encounter with building departments can be attributed to not submitting a complete application with the necessary construction documents, not hiring competent contractors, or undertaking a project themselves without the necessary skills. By using competent professionals to draft the plans, submitting all required documents and information to the building department, and responding promptly to concerns raised during plan review, you will shorten the waiting time for permit approval.

There are still potential problems that may occur during construction, but these are avoidable by following the proper procedure. Always get approval for changes to the construction documents, call for inspections when required, and fix problems when

Figure 14–2   It is important to hire competent contractors.

required. If you don't agree with an order of the building official, appeal the order, don't ignore it.

If you do run afoul of the building code and receive a notice of violation or a stop work order, determine immediately what you need to do to comply with the building official's order so the building process isn't delayed. You can avoid court or an administrative hearing by communicating with the building official and demonstrating a good-faith effort to comply. If you have established a sterling reputation in your prior dealings with the building inspector for honesty and competency, he will work with you to resolve any difficulties.

Figure 14–3   Invest in your relationship with the building inspector—practice good communication and honesty. Your inspector will be more willing to help you in return.

Hopefully you will never have to deal with corrupt government employees or incompetent ones, but, if you do, go to the proper authorities who can assist you.

You have a lot of control over how smoothly your building project proceeds in your dealings with local government. By following the suggestions set forth in the earlier chapters of this book, you can avoid 99% of the difficulties encountered by other people who are not as conscientious as they ought to be.

Undertaking a construction project can be an exciting experience as you contemplate the end result. By taking the necessary steps discussed in these chapters, you will sail through the permit process and avoid costly delays from the time you begin to contemplate the project until it is ready for occupancy.

GLOSSARY OF TERMS

abatement   An action taken by a building official to stop an unlawful act.

accessory use   A structure and/or use that is subordinate to and serves a principal structure or use.

administrative hearing   A judicial proceeding that is allowed by state law that takes place within the local jurisdiction, not in a courtroom.

admission   A statement by a person that can be used against him because it admits elements of the offense.

appeal   A procedure whereby a person who has received a notice of violation can contest the decision of the building official. Under the building code, an appeal is heard before the Board of Appeals.

arraignment   The formal act of calling the defendant into open court, informing him or her of the offense with which he or she is charged, and asking him or her whether he or she is guilty or not guilty.

bench trial   A trial in which the judge is the trier of fact.

Board of Appeals   A group of people appointed by the local governing body to hear appeals from decisions of the building official.

boundary   The extent of the ownership of land on a parcel of property.

bribery   An illegal offering of something of value in return for the performance or failure to perform an action involving a governmental official.

building code   A comprehensive collection of regulations that determine the manner in which structures are constructed, altered, repaired, occupied, or demolished that is adopted by local governments as the law governing those activities.

building inspector   A person working on behalf of a local government who reviews building construction.

building line   The line established by law, beyond which a building shall not extend, except as specifically provided by law. (*2006 International Building Code*)

building official   A person appointed by local government to oversee the building department. A building official has many powers to regulate construction in a jurisdiction.

building permit application   A written request that an official document be issued by the authority having jurisdiction giving permission to a particular person or entity for the performance of a specific activity.

burden of proof   A legal concept that describes who must prove the case in court by submitting enough evidence. The burden of proof is usually on the party bringing the court action.

certificate of occupancy   A type of permit issued by a local government that allows the use and occupancy or a change in the existing occupancy classification of a building or structure. It is not supposed to be issued until the building or structure meets the conditions of the building code and after a final inspection is performed.

chancery court   A court that has special equitable powers (e.g., to order the demolition of property).

code   A body of laws adopted by a unit of local government.

complainant   A person who signs a complaint charging a defendant with a violation.

complaint   The charging document that is the basis for an action against a defendant in a court of law or administrative hearing. Other names for the same document are **ticket** or **citation.** In some jurisdictions, **notice of violation** is used.

compliance   An action by a party that abates a problem with a building or structure.

construction documents   The written, graphic, and pictorial documents prepared or assembled for describing the design, location, and physical characteristics of the elements of the project necessary for obtaining a permit. (*2006 International Building Code*)

corporation   A legal entity authorized by state law that consists of shareholders, a board of directors, and officers. It can own property and incur debt.

covenant   An agreement or promise between two or more parties. A covenant that runs with the land goes with the land, cannot be separated from it or transferred without it. (*Black's Law Dictionary*)

defendant   A person who is prosecuted by a local jurisdiction or the state.

due process of law   The procedure that every person is entitled to before losing his liberty or property. The law of the land, being the U.S. Constitution, state law, and local law, must be followed. It requires that a statute give fair warning of conduct that it prohibits so a person can avoid violating the law. Before someone can be fined or put in jail, due process demands that the person affected has the right to notice and an opportunity to be heard before the tribunal that will decide the issue. It includes the right to present evidence in one's defense.

easement   A right of the owner of one parcel of land, by reason of such ownership to use the land of another for a special purpose not inconsistent with a general property in the owner. (*Black's Law Dictionary*)

enforcement   Actions taken by a building official to ensure that the building code is followed by persons who are not willingly following the code and the orders of the building official. Enforcement actions include notifying the offender of a violation, demanding compliance, issuing a stop work order, revoking a certificate of occupancy, and filing a charging document against the offender.

evidence   Statements and items that are introduced at a trial or administrative hearing to prove the elements of the offense and the identity of the responsible party.

examination   The series of questions put to a witness by a party to the action or her counsel, for the purpose of bringing before the court and jury in legal form the knowledge that the witness has of the facts and matters in dispute, or of probing and sifting the evidence previously given. (*Black's Law Dictionary*) **Direct examination** is the examination of one's own witnesses. **Cross-examination** is the examination of opposing witnesses.

exempt work   Activity that does not require a permit.

fire code   A comprehensive collection of regulations that affect structures to reduce the possibility of hazard from fire or explosion.

flood hazard area   An area designated by a governmental authority as being particularly at risk for flooding.

grading permit   An official document or certificate issued by the authority having jurisdiction that authorizes activity so water flows and drains properly throughout the neighborhood.

hearsay   Evidence that seeks to prove the truth of the matter asserted. Hearsay evidence is generally not admissible unless it falls within an exception to the rule against hearsay.

home occupation   A business that is conducted within a residential structure in a residential zoning district.

inspection   An examination or search of property by a building official or inspector.

jury trial   A trial in which 6 or 12 citizens are the triers of fact.

land trust   A legal way of holding title to land with actual legal ownership vested in a trustee and with the beneficiaries getting the benefit of the property.

legal authority   Power given by the law, such as ownership of property.

legal nonconforming use   A structure and the use thereof or the use of land that does not comply with the regulations of the title governing use in the district in which it is located, but which conformed with all of the codes, ordinances, and other legal requirements applicable at the time such structure was erected, enlarged, or altered, and the use thereof or the use of land was established.

legal responsibility   Liability for violations that occur on property.

license   A permit from a governmental body allowing someone to perform a certain function (e.g., contracting, plumbing, or roofing).

limited liability company (LLC)   An unincorporated entity organized under state law that affords limited liability to its owners. Often run by a manager under an operating agreement.

local jurisdiction   A unit of local government such as a municipality or county.

model code   A sample code developed by various private organizations dealing with a specific topic that is adopted by local governments as their law. Building codes are usually based on model codes.

notice of violation   A written document served on a responsible party that contains an order by the building official requiring that corrections of violations of the building code be made.

objections   The way in which parties in a lawsuit challenge the introduction of evidence.

occupancy classification   A regulation that sets forth how a building can be used by particular people (e.g., assembly, business, group facilities, residential).

offense   A violation of the building code.

ordinance   A law passed by a unit of local government such as a municipality or county.

permit   An official document or certificate issued by the authority having jurisdiction that authorizes performance of a specified activity.

plea bargaining   A conference between the prosecutor and the defendant or the defendant's attorney during which the parties try to reach an agreement as to the ultimate disposition of the case.

proof of service   A document signed by a building official that identifies the type of service used to serve a responsible party with a notice of violation or other document.

Property Index Number (PIN)   A unique number assigned to a parcel of property in order to identify it. It is typically used by a tax assessor's office.

prosecution   A legal proceeding in which a person's liability for a violation of the law is determined.

prosecutor   The attorney representing the unit of local government in an enforcement action.

registered agent   A person, including a corporation, authorized by law to receive notice and service of process on behalf of a corporation or limited liability company.

required inspections   Those inspections that are mandated by the building code.

responsible party   The person or legal entity (e.g., corporation, limited liability company) who is responsible under the building code for the condition of the property, structure, or equipment at issue. All notices and complaints are directed to the responsible party.

right of entry   The authority of the building official to enter private property and conduct inspections, and the restrictions on that power imposed by the Fourth Amendment of the Bill of Rights of the U.S. Constitution.

scope   The extent to which a law applies to a particular action.

sentencing   That point in the court proceeding after a defendant is found guilty during which the judge imposes the penalty against the offender.

setback   The specific number of feet set forth in the local zoning code in which no building may be placed.

special or conditional use   A use allowed under the zoning code that may be allowed in a district only with the consent of the zoning board.

statute   A law passed by the legislature of a state.

statute of limitations   A law that limits the length of time during which a defendant can be prosecuted for a violation of a statute or ordinance.

stop work order   An order by the building official that work on a property or structure cease.

stormwater management   Regulations and statutes adopted by the state and local government to reduce the risk of damage to persons and property in the event of a flood.

survey   To determine and delineate the form, extent, and position (as of a tract of land) by taking linear and angular measurements and by applying the principles of geometry and trigonometry. (*Merriam-Webster's Online Dictionary*)

temporary certificate of occupancy   A type of permit issued by a local government that allows the temporary use and occupancy or a change in the existing occupancy classification of a building or structure as long as specific requirements are met.

trial   A judicial examination, in accordance with law of the land, of a cause, either civil or criminal, of the issues between the parties, whether of law or fact, before a court that has jurisdiction over it. (*Black's Law Dictionary*)

unlawful act   A violation of the building code.

use   The purpose or activity for which the land or building thereon is designed, arranged, or intended or for which it is occupied or maintained.

zoning   The division of a city by legislative regulation into districts and the prescription and application in each district of regulations having to do with structural and architectural designs of buildings and of regulations prescribing use to which buildings within designated districts may be put. (*Black's Law Dictionary*)

APPENDIX

Sample Model Code Provisions

PERMITS

2006 International Building Code

Section 105.1 Required. Any owner or authorized agent who intends to construct, enlarge, alter, repair, move, demolish, or change the occupancy of a building or structure, or to erect, install, enlarge, alter, repair, remove, convert or replace any electrical, gas, mechanical or plumbing system, the installation of which is regulated by this code, or to cause any such work to be done, shall first make application to the building official and obtain the required permit.

2006 International Residential Code

Section R105.1 Required. Any owner or authorized agent who intends to construct, enlarge, alter, repair, move, demolish, or change the occupancy of a building or structure, or to erect, install, enlarge, alter, repair, remove, convert or replace any electrical, gas, mechanical or plumbing system, the installation of which is regulated by this code, or to cause any such work to be done, shall first make application to the building official and obtain the required permit.

2006 International Fire Code

Section 105.1 Permits required. Permits required by this code shall be obtained from the fire code official. Permit fees, if any, shall be paid prior to issuance of the permit. Issued permits shall be kept on the premises designated therein at all times and shall be readily available for inspection by the fire code official.

INSPECTIONS

2006 International Building Code

Section 109.1 General. Construction or work for which a permit is required shall be subject to inspection by the building official and such construction or work shall remain accessible and exposed for inspection purposes until approved.

2006 International Residential Code

Section 109.1 Types of inspections. For onsite construction, from time to time the building official, upon notification from the permit holder or his agent, shall make or cause to be made any necessary inspections and shall either approve that portion of the construction as completed or shall notify the permit holder of his or her agent wherein the same fails to comply with this code.

2006 International Fire Code

Section 106.2 Inspections. The fire code official is authorized to conduct such inspections as are deemed necessary to determine the extent of compliance with the provisions of this code and to approve reports of inspection by approved agencies or individuals.

CERTIFICATES OF OCCUPANCY

2006 International Building Code

Section 110.1 Use and occupancy. No building or structure shall be used or occupied, and no change in the existing occupancy classification of a building or structure or portion thereof shall be made until the building official has issued a certificate of occupancy therefore as provided herein. Issuance of a certificate of occupancy shall not be construed as an approval of a violation of the provisions of this code or of other ordinances of the jurisdiction.

2006 International Residential Code

Section R110.1 Use and occupancy. No building or structure shall be used or occupied, and no change in the existing occupancy classification of a building or structure or portion thereof shall be made until the building official has issued a certificate of occupancy therefore as provided herein. Issuance of a certificate of occupancy shall not be construed as an approval of a violation of the provisions of this code or of other ordinances of the jurisdiction.

Certificates presuming to give authority to violate or cancel the provisions of this code or other ordinances of the jurisdiction shall not be valid.

2006 International Fire Code

Section 105.3.3 Occupancy prohibited before approval. The building or structure shall not be occupied prior to the fire code official issuing a permit that indicates that applicable provisions of the code have been met.

VIOLATIONS

2006 International Building Code

Section 113.1 Unlawful acts. It shall be unlawful for any person, firm or corporation to erect, construct, alter, extend, repair, move, remove, demolish or occupy any building, structure or equipment regulated by this code, or cause same to be done in conflict with or in violation of any of the provisions of this code.

2006 International Residential Code

Section R113.1 Unlawful acts. It shall be unlawful for any person, firm or corporation to erect, construct, alter, extend, repair, move, remove, demolish or occupy any building, structure or equipment regulated by this code, or cause same to be done in conflict with or in violation of any of the provisions of this code.

2006 International Fire Code

Section 109.1 Unlawful acts. It shall be unlawful for any person, firm or corporation to erect, construct, alter, extend, repair, remove, demolish or utilize a building, occupancy, premises or system regulated by this code, or cause same to be done in conflict with or in violation of any of the provisions of this code.

INDEX